Robert E. Gough, PhD
Ronald F. Korcak, PhD
Editors

Blueberries:
A Century of Research

Pre-publication
REVIEWS,
COMMENTARIES,
EVALUATIONS . . .

"**T**his conference proceedings gives a fascinating update of current research in blueberries. Several of the chapters give reviews of the great strides the breeders have made in the last one hundred years. Others document the current efforts being made to adapt blueberries for higher pH mineral soils."

Adam Dale, PhD
Research Scientist, Berry Crops,
Horticultural Research Institute
of Ontario

" **A** number of prominent researchers working with blueberries participated in the 7th North American Blueberry Research-Extension Workers Conference in July 1994. Their reports are assembled in this book, *Blueberries: A Century of Research*, edited by Robert Gough and Ronald Korcak. . . . For nearly a century, plant breeders have been selecting highbush blueberry germplasm from the field and, through cross-pollination and hybridization, developed cultivars bearing desirable attributes for growth, bloom, berry qualities and disease resistance. . . . (H)ighbush blueberry cultivars are increasingly derived from crosses between a number of genetically compatible species. Several opening chapters concisely summarize the history of some of these blueberry breeding programs. Several subsequent and authoritative chapters then introduce the results of creative new breeding programs that involve native deerberry (*V. stamineum*) and sparkleberry (*V. arboreum*). . . . Additional chapters address aspects of disease and pest detection and identification, and breeding for resistance in commercial blueberries. . . . Finally informed growers will find considerable new practical information to aid them in commercial production with highbush blueberries. Topics include soil nutrient availability, disease and pests detection, mulching and hormonal fruit induction."

James H. Cane, PhD
Associate Professor
Auburn University

Blueberries:
A Century of Research

Blueberries:
A Century of Research

Robert E. Gough, PhD
Ronald F. Korcak, PhD
Editors

Food Products Press
An Imprint of
The Haworth Press, Inc.
New York · London

Published by

Food Products Press, 10 Alice Street, Binghamton, NY 13904-1580 USA

Food Products Press is an imprint of The Haworth Press, Inc., 10 Alice Street, Binghamton, NY 13904-1580 USA.

Blueberries: A Century of Research has also been published as *Journal of Small Fruit & Viticulture*, Volume 3, Numbers 2/3 and 4 1995.

Library of Congress Cataloging-in-Publication Data

Blueberries : a century of research / Robert E. Gough, Ronald F. Korcak, editors.
     p. cm.
    "Proceedings from the 7th North American Blueberry Research-Extension Workers Conference, July 5-8, 1994, Beltsville, Maryland"–Prelim. p.
    "Has also been published as Journal of small fruit & viticulture, volume 3, numbers 2/3 and 4 1995"–T.p. verso.
    Includes bibliographical references.
    ISBN 1-56022-053-8 (alk. paper)
    1. Blueberries Congresses. I. Gough, Robert E. (Robert Edward) II. Korcak, Ronald F. III. North American Blueberry Research-Extension Workers Conference (7th : 1994 : Beltsville, MD)
SB386.B7B58 1996   96-1129
634'.737–dc20   CIP

INDEXING & ABSTRACTING

Contributions to this publication are selectively indexed or abstracted in print, electronic, online, or CD-ROM version(s) of the reference tools and information services listed below. This list is current as of the copyright date of this publication. See the end of this section for additional notes.

- *Abstracts on Rural Development in the Tropics (RURAL),* Royal Tropical Institute (KIT), 63 Mauritskade, 1092 AD Amsterdam, The Netherlands

- *AGRICOLA Database, National Agricultural Library,* 10301 Baltimore Boulevard, Room 002, Beltsville, MD 20705

- *BIOBUSINESS:* covers business literature related to the life sciences; covers both business & life science periodicals in such areas as pharmacology, health care, biotechnology, foods & beverages, etc. BIOSIS, Bibliographic Control Department, 2100 Arch Street, Philadelphia, PA 19103-1399

- *Biosciences Information Service of Biological Abstracts (BIOSIS),* Biosciences Information Service, 2100 Arch Street, Philadelphia, PA 19103-1399

- *Cambridge Scientific Abstracts,* Microbiology Abstracts Section, Environmental Routenet (accessed via INTERNET), 7200 Wisconsin Avenue #601, Bethesda, MD 20814

- *CNPIEC Reference Guide: Chinese National Directory of Foreign Periodicals,* P.O. Box 88, Beijing, Peoples Republic of China

- *Foods Adlibra,* Foods Adlibra Publications, 9000 Plymouth Avenue North, Minneapolis, MN 55427

- *Food Intelligence on Compact Disc* (covers the food industry, including journals in business, foodservice management, aquaculture, biochemistry, food preparation, irradiation, microbiology, nutrition and toxocology). BIOSIS, Bibliographic Department, 2100 Arch Street, Philadelphia, PA 19103-1399

(continued)

- *Food Science and Technology Abstracts (FSTA),* scanned, abstracted and indexed by the International Food Information Service (IFIS) for inclusion in Food Science and Technology Abstracts (FSTA), International Food Information Service, Lane End House, Shinfield, Reading RG2 9BB, England

- *GEO Abstracts (GEO Abstracts/GEOBASE),* Elsevier/GEO Abstracts, Regency House, 34 Duke Street, Norwich NR3 3AP, England

- *Horticultural Abstracts (HA/CAB ABSTRACTS),* c/o CAB International/CAB ACCESS . . . available in print, diskettes updated weekly, and on INTERNET. Providing full bibliographic listings, author affiliation, augmented keyword searching. CAB International, P.O. Box 100, Wallingford Oxon OX10 8DE, UK

- *INTERNET ACCESS (& additional networks) Bulletin Board for Libraries ("BUBL"), coverage of information resources on INTERNET, JANET, and other networks.*
  - JANET X.29: UK.AC.BATH.BUBL or 00006012101300
  - TELNET: BUBL.BATH.AC.UK or 138.38.32,45 login 'bubl'
  - Gopher: BUBL.BATH.AC.UK (138.32.32.45). Port 7070
  - World Wide Web: http://www.bubl.bath.ac.uk./BUBL/home.html
  - NISSWAIS: telnetniss.ac.uk (for the NISS gateway)
  The Andersonian Library, Curran Building, 101 St. James Road, Glasgow G4 ONS, Scotland

- *Referativnyi Zhurnal (Abstracts Journal of the Insitute of Scientific Information of the Republic of Russia),* The Institute of Scientific Information, Baltijskaja ul., 14, Moscow A-219, Republic of Russia

- *Viticulture and Enology Abstracts (VITIS),* Bundesanstalt fur Zuchtungsforschung an Kulturpflanzen, Institut fur Rebenzuchtung Geilweilerhof, D-76833 Siebeldingen, Federal Republic of Germany

(continued)

SPECIAL BIBLIOGRAPHIC NOTES

related to special journal issues (separates)
and indexing/abstracting

☐ indexing/abstracting services in this list will also cover material in any "separate" that is co-published simultaneously with Haworth's special thematic journal issue or DocuSerial. Indexing/abstracting usually covers material at the article/chapter level.

☐ monographic co-editions are intended for either non-subscribers or libraries which intend to purchase a second copy for their circulating collections.

☐ monographic co-editions are reported to all jobbers/wholesalers/approval plans. The source journal is listed as the "series" to assist the prevention of duplicate purchasing in the same manner utilized for books-in-series.

☐ to facilitate user/access services all indexing/abstracting services are encouraged to utilize the co-indexing entry note indicated at the bottom of the first page of each article/chapter/contribution.

☐ this is intended to assist a library user of any reference tool (whether print, electronic, online, or CD-ROM) to locate the monographic version if the library has purchased this version but not a subscription to the source journal.

☐ individual articles/chapters in any Haworth publication are also available through the Haworth Document Delivery Services (HDDS).

Proceedings
from the 7th North American Blueberry Research-Extension Workers Conference

July 5-8, 1994
Beltsville, Maryland

Blueberries:
A Century of Research

CONTENTS

RELATED POSTERS

SESSION IV: BLUEBERRY CULTURE
      Moderator: Nicholi Vorsa

RELATED POSTERS

ABOUT THE EDITORS

Robert E. Gough, PhD, is the former President of the Northeast Region of the American Society for Horticultural Science and a leading specialist in small fruit and viticulture. Dr. Gough is now Associate Professor of Horticulture at Montana State University in Bozeman. He has a diverse background, having earned a BA in English, an MS in Horticulture, and a PhD in Botany. His experiences as a county agricultural agent, state and regional extension specialist in small fruit, and a senior research scientist have provided him with a great deal of insight into the needs of growers.

Dr. Gough has published extensively–nearly 300 articles–in the area of pomology (fruit science) in both scientific journals and popular magazines, including the *Journal of the American Society for Horticultural Science, HortScience, Scientia Horticulturae, Journal of Small Fruit & Viticulture, New England Gardener, Harrowsmith/ Country Life, Country Journal, Fine Gardening, National Gardening, New England Farmer, American Fruit Grower, Proceedings of the New England Small Fruit Conference,* and *Proceedings of the Ohio Fruit Congress.* He also maintains an active interest in all areas relating to general horticulture, agriculture, crop science, soil science, and community gardening/landscape architecture. A member of numerous professional and honorary societies, Dr. Gough served as an Associate Editor for the *Journal of the American Society for Horticultural Science* and *HortScience* from 1985-1988. Currently, he is Editor of *Journal of Small Fruit & Viticulture,* Senior Editor for horticulture, and a consultant for Food Products Press imprint.

Ronald F. Korcak, PhD, is Research Leader of the Fruit Laboratory at the Agricultural Research Service Center in Beltsville, Maryland. He has over 17 years of experience in the nutrition of fruits and has worked in the field of by-product utilization in agriculture.

Foreword

The 7th North American Blueberry Research-Extension Workers Conference, jointly sponsored by the Fruit Laboratory, Beltsville, Maryland and the Blueberry/Cranberry Research Station, Chatsworth, New Jersey, was held at the Beltsville Agricultural Research Center, 5-8 July, 1994. Participants toured a Maryland pick-your-own blueberry operation before the conference and spent all day touring the Blueberry/Cranberry Research Station in Chatsworth, New Jersey between two days of oral and poster sessions.

This conference continued the tradition of the National Blueberry Conference held at Rutgers University, 6-7 July, 1960, to celebrate a half-century of blueberry research that commenced with the publication of Dr. F.V. Coville's *Experiments in Blueberry Culture* (United States Department of Agriculture Bureau of Plant Industry Bulletin 193 [1910]). Most of the participants at that first conference have since retired or passed away, though some, including Drs. Gene Galletta, Paul Eck, Miklos Faust, Alan Stretch, Walt Kender, Warren Stiles, and Jim Moore (Chairman of the first conference) continue to make important contributions to fruit research. Subsequent conferences were sponsored by the University of Maine (6-7 April, 1966), Michigan State University (6-7 November, 1974), University of Arkansas (16-18 October, 1979), University of Florida (1-3 February, 1984), and Washington State University (10-12 July, 1990).

While the conference traditionally has included research reports contributed primarily from the United States, Canadian researchers have maintained a strong presence, reporting on their findings first

[Haworth co-indexing entry note]: "Foreword." Gough, Robert E., and Ronald F. Korcak. Co-published simultaneously in *Journal of Small Fruit & Viticulture* (Food Products Press, an imprint of The Haworth Press, Inc.) Vol. 3, No. 2/3, 1995, pp. xvii-xviii; and: *Blueberries: A Century of Research* (ed: Robert E. Gough, and Ronald F. Korcak) Food Products Press, an imprint of The Haworth Press, Inc., 1995, pp. xvii-xviii. Single or multiple copies of this article are available from The Haworth Document Delivery Service [1-800-342-9678, 9:00 a.m. - 5:00 p.m. (EST)].

xvii

at the second conference and at each of the subsequent conferences. Polish researchers have been involved since the fourth conference, and this year we heard the first report from our Japanese colleagues. Future conferences will no doubt include papers by researchers from additional countries as the blueberry becomes more and more popular around the world.

We thank all who participated in this conference and who submitted their research reports for publication in this volume. The 8th North American Blueberry Research-Extension Workers Conference will be held at Castle Hayne, North Carolina in 1998.

Robert E. Gough
Ronald F. Korcak

SESSION I:
BLUEBERRIES OF THE FUTURE
Moderator: Ronald F. Korcak

Utilization of Wild Blueberry Germplasm: The Legacy of Arlen Draper

J. F. Hancock
W. A. Erb
B. L. Goulart
J. C. Scheerens

SUMMARY. The native North American blueberry species have a very broad ecological amplitude. Dr. Arlen Draper of the USDA-

J. F. Hancock is Professor, Department of Horticulture, Michigan State University, East Lansing, MI 48824.

W. A. Erb is Associate Professor, Kansas State University Horticulture Center, Wichita, KS 67233.

B. L. Goulart is Associate Professor, Department of Horticulture, Pennsylvania State University, University Park, PA 16802.

J. C. Scheerens is Associate Professor, Department of Horticulture, Ohio Agricultural Research and Developmental Center, Wooster, OH 44691.

[Haworth co-indexing entry note]: "Utilization of Wild Blueberry Germplasm: The Legacy of Arlen Draper." Hancock, J. F. et al. Co-published simultaneously in *Journal of Small Fruit & Viticulture* (Food Products Press, an imprint of The Haworth Press, Inc.) Vol. 3, No. 2/3, 1995, pp. 1-16; and: *Blueberries: A Century of Research* (ed: Robert E. Gough, and Ronald F. Korcak) Food Products Press, an imprint of The Haworth Press, Inc., 1995, pp. 1-16. Single or multiple copies of this article are available from The Haworth Document Delivery Service [1-800-342-9678, 9:00 a.m. - 5:00 p.m. (EST)].

1

Beltsville has made substantial progress in highbush blueberry breeding through the liberal use of this genetic gold mine. He has incorporated the genes of at least 10 species into the genetic background of the highbush blueberry and several of his creations have been released as cultivars. His hybrid US 75 (*V. darrowi* × *V. corymbosum*) has played a critical role in the development of the southern highbush types with a low chilling requirement such as 'Cooper,' 'Georgiagem,' 'Gulfcoast,' and 'O'Neal.' His most complex release to date, 'Sierra,' has the genes of 5 species and appeared to be winter hardy. Other unreleased hybrids should provide breeders with a diverse combination of valuable characteristics such as tolerance to mineral soils, disease and pest resistance, drought tolerance and fruit quality. His inter-species hybrids can be used in both northern and southern breeding programs, as winter hardy types have been recovered from backcross populations that contain high percentages of southern derived genes. *[Article copies available from The Haworth Document Delivery Service: 1-800-342-9678.]*

KEYWORDS: *Vaccinium*; Wide species crosses; Cold tolerance

INTRODUCTION

A wealth of germplasm is literally at the fingertips of North American blueberry breeders. A wide array of native species are found across a broad range of natural environments and most of these species can be hybridized to at least some extent with cultivated types (Ballington, 1990; Lyrene and Ballington, 1986). This storehouse of genetic variability affords breeders with many exciting opportunities.

Vaccinium breeding is a very recent development (Eck and Childers, 1966). Highbush breeding began in the early twentieth century in New Jersey, with the first hybrids being released by Dr. Frederick Coville of the United States Department of Agriculture (USDA). Rabbiteye breeding was initiated in the 1940s and the first widespread cultivars were made available in the 1950s. Lowbush breeding projects have existed for only a few decades and hybrid cultivars have not been widely planted.

The genetic variability found within *Vaccinium corymbosum* L. has made breeding progress rapid, but of increasing importance is the use of wild populations of compatible species (Galletta, 1975;

Ballington, 1990). Thoughtful breeders like Dr. Arlen Draper of USDA have made liberal use of this genetic gold mine to expand the range of blueberry cultivation and develop unique adaptations. It is the object of this article to discuss the wild species and their patterns of variability, and honor the brilliant effort of Draper to exploit this germplasm resource. Draper recognized early in his career that most of the native species of blueberries could be hybridized with cultivated types and provide unique genes (Draper, 1977; Draper et al., 1982). Many other breeders have utilized native species in their blueberry improvement programs (Moore, 1975; Sharp and Sherman, 1971; Ballington, 1990; Lyrene, 1990), but no other program has matched the scope of Draper's. He has now incorporated the genes of at least 10 species into the genetic background of the highbush blueberry and several of his genetic masterpieces have been released as cultivars (Hancock and Goulart, 1993; Ehlenfeldt and Vorsa, 1994).

The Species Situation

Commercial blueberries belong to section *Cyanococcus* of *Vaccinium*. Three species are of major economic importance: (1) the highbush blueberry, *V. corymbosum* L., (2) the lowbush blueberry, *V. angustifolium* Aiton, and (3) the rabbiteye blueberry, *V. ashei* Reade. Most of the worldwide blueberry production comes from the highbush blueberry, although lowbush and rabbiteye types are important in the northeastern and southwestern regions of North America (Hanson and Hancock, 1990).

Blueberries exist at three ploidy levels 2x (2n = 24), 4x (2n = 48) and 6x (2n = 72) (Luby et al., 1991; Vander Kloet, 1988). The native species are scattered over a wide array of environments in North America and have a number of potentially useful horticultural traits (Luby et al., 1991). They range from the bitterly cold regions of northeastern Canada to the winter tourist zones of Florida (Figure 1). All are acidophiles, but some have much broader soil adaptations than others. The diploid species are found in series along the eastern part of North America from Florida to the maritime provinces. The hexaploids are concentrated in the southeastern portion of the United States, while tetraploid populations of the

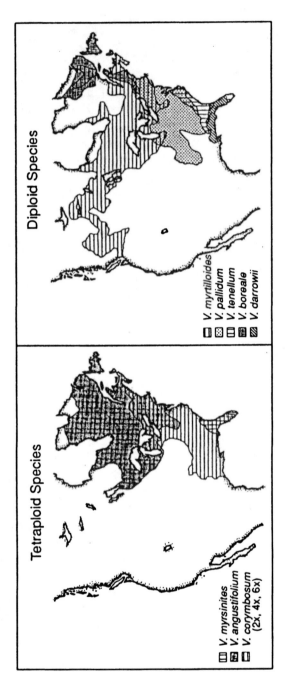

FIGURE 1. Geographic range of North American blueberry species. While it is taxonomically correct to join diploid, tetraploid and hexaploid highbush blueberries into *V. corymbosum* (Vander Kloet, 1980), many breeders follow Camp's older taxonomy which separates the diploid populations into *V. atrococcum* and *V. elliottii*, and hexaploid populations into *V. ashei* and *V. constablaei* (Luby et al., 1991). The diploid and hexaploid populations are concentrated in the southeastern US.

4

most widely distributed species, *V. corymbosum*, range from southern Canada to east Texas and Florida.

Native *V. corymbosum* and *V. ashei* are compilospecies composed of a very complex background (Hancock, 1993; Vander Kloet, 1988). Vander Kloet (1980) found the variation patterns in highbush and rabbiteye species to be so complex that he placed all the crown forming, *Cyanococcus* taxa (2x, 4x and 6x) under one name, *V. corymbosum*. He combined several species still recognized by plant breeders because of their unique characteristics, including the diploids *V. atrococcum* and *V. elliottii*, and the hexaploids–*V. ashei* and *V. contstablaei* (Luby et al., 1991).

Several southern US races of diploid *Vaccinium* probably combined to form tetraploid *V. corymbosum*, and this species continues to interact with numerous other *Vaccinium* species wherever their ranges overlap (Vander Kloet, 1988). Hexaploid *V. corymbosum* (*V. ashei*) may also have arisen in the southeastern US from repeated hybridizations of several diploid and tetraploid species. *Vaccinium angustifolium* appears to be a direct descendant of *V. pallidum* × *V. boreale* (Hall and Aalders, 1961) but introgression with *V. corymbosum* may have also influenced its subsequent development (Vander Kloet, 1988).

The most astonishing characteristic of the blueberry species is their extreme promiscuity. There is little electrophoretic or cytogenetic variation among species (Vorsa et al., 1988; Bruederle, 1991), and *V. corymbosum* has been shown to have tetrasomic inheritance indicating it is an autotetraploid (Draper and Scott, 1971; Krebs and Hancock, 1989). The tetraploids, *V. corymbosum*, *V. angustifolium* and *V. myrsinites* Lamarck, can be readily hybridized to produce fertile F_1 progeny (Darrow and Camp, 1945; Draper, 1977), even though *V. myrsinites* has a distinct ecological range from the other two species. Hybrids of the highbush and lowbush are relatively common where their habitats overlap. Reproductive isolating barriers are more developed in the diploid taxa, but all of the species can be crossed to some degree and produce fertile hybrids (Ballington and Galletta, 1978; Lyrene and Ballington, 1986).

Fertile hybrids can also be produced between most of the diploid species and the polyploids (Darrow et al., 1954; Draper, 1977). Diploids can be readily crossed with tetraploids via unreduced gametes (Lyrene and Sherman, 1983; Mengalos and Ballington, 1988; Ortiz et al., 1992), and crosses between hexaploid *V. ashei* and the diploid and tetraploid species have been successful, although fertility is reduced due to chromosomal imbalances (Darrow et al., 1949; Goldy and Lyrene, 1984). The hexaploids and tetraploids have also been successfully intermated to produce partially fertile F_1 hybrids (Jelenkovic and Draper, 1973; Vorsa, 1987, 1988; Vorsa et al., 1987).

DRAPER'S USE OF WILD GERMPLASM

Working with D. H. Scott and G. J. Galletta, Draper used a number of different strategies to incorporate the genes of native species into the cultivated highbush background. He produced interspecific tetraploid hybrids of *V. myrsinites* Michaux × *V. angustifolium* that could be directly crossed with cultivated highbush types (JU 60, 62, 64). He colchicine-doubled diploid hybrids of *V. myrtilloides* × *V. corymbosum* (US 226) and then crossed this hybrid with tetraploid *V. corymbosum*. He also generated numerous hybrids of diploid and polyploid species and the most fertile ones were used as breeding parents (JU 10, JU 11, US 17, US 75, US 79). The cross of *V. darrowi* Camp × *V. corymbosum* yielded a tetraploid hybrid (US 75) that proved to be completely interfertile with the highbush types, while the other crosses yielded less fertile pentaploids (US 79) or hexaploids (NJUS 10 and 11). Fertility in the complex hybrids ranged from low to moderate, but all have been successfully crossed with highbush types (Draper, 1977; Draper et al., 1982).

After the initial complex hybrids were generated, Draper began to backcross them into the highbush background. The hybrid US 75 has proven to be the most useful parent to date, as it has taken only two backcross generations to develop a wide array of low chill cultivars for the southern US (Ballington, 1990). Several southern breeders have worked closely with Draper and used US 75 as a parent, including Max Austin at the University of Georgia, Gene Galletta and Jim Ballington at North Carolina State, Jim Spears at the USDA Mississippi, and Jim Moore at the University of Arkansas. A similar strategy has been used independently by Sharp, Sherman and Lyrene to produce low chill cultivars for Florida (Sharp and Sherman, 1971; Lyrene, 1990).

Hybrids with *Vaccinium angustifolium* have also proven to be very useful in breeding highbush blueberries. Within a few backcross generations, Draper was able to use hybrids developed by his predecessors to generate several early blooming cultivars with *V. angustifolium* in their background ('Sunrise' and 'Patriot'). Likewise, Jim Luby and Cecil Stushnoff at the University of Minnesota and Stanley Johnston at Michigan State University used the same strategy to produce cultivars with increased cold hardiness (Finn et al., 1990).

In addition to backcrossing, Draper mixed the genes of species by intercrossing elite hybrids. His most complex hybrids now carry 6 species in their background (Table 1) and he continues to push forward. Amazingly, most of these lines have the habit of highbush types and are highly fertile, even though many are less than 50% *V. corymbosum*. Most of these mixtures are not yet of cultivar quality; however, 'Sierra' (G640) was recently named with the genes of 5 species (Table 1).

THE FUTURE

The overall potential of the Draper material has only begun to be recognized. While most of the attention has been on fruit characteristics and chilling hours, there is a whole array of other useful traits contained in this germplasm that has not yet been exploited. The wide ecological range of the native species material should provide a diverse combination of valuable characteristics, thus increasing the available land for successful blueberry production.

While *Vaccinium darrowi* has been used primarily to reduce the chilling requirement of the highbush blueberry, it may also play a key role in increasing the soil and temperature adaptations of *V. corymbosum*. Hybrids of *V. darrowi* have a much higher photosynthetic heat tolerance than the highbush blueberry (Moon et al., 1987a and b; Hancock et al., 1992) and they have greater drought tolerance and adaptation to mineral soils (Erb et al., 1988 and 1993; Korcak, 1986a, 1986b, 1989a; Korcak et al., 1982). While these

TABLE 1. Selected complex hybrids of Arlen Draper that show potential as breeding parents. Characteristics: 1-yield (+ high, − low), 2- vigor (+ high, − low), 3-fruit color (+ powderblue, − black), 4-fruit size (+ Large, − small), 5-scar (+ small, − large), 6-firmness (+ firm, − soft), 7-flavor (+ good, − poor).

Number	Cross	Species Background	Strengths and Weaknesses
G245	US75 × G100	70 cor^z/25 dar/5 ang	3+,5+,6+
G344	US75 × Eliz	73 cor/25 dar/2 ang	3+,5+,6+,7+
US612	G362 × JU64	53 cor/28 ang/25 mys	3−,6+
US621	G362 × US226	47 cor/25 atr/3 ang/25 myt	1+,3−,6+
US636	US75 × US226	23 cor/25 dar/25 atr/2 ang/25 myt	1+,2+,3−
US643			3+,6+,7+
US647	US75 × G362	70 cor/25 dar/5 ang	3+,4+,6−
US654			3+
US665	G362 × US75		3+,6+,7+
US667			3+
US671			3+
US673	G362 × JU11	47 cor/20 atr/30 ash/3 ang	2+
US676			1+,2+,3+
US696	JU11 × US226	45 atr/30 ash/25 myt	2+,3−
US702	G362 × JU64	47 cor/28 ang/25 mys	2+,3−
US717	JU64 × G362	47 cor/28 ang/25 mys	2+,3−
US720	JU11 × US75	23 cor/25 dar/20 atr/30 ash/2 ang	2+,3−
US723			2+,3−
US729	US75 × JU11	23 cor/25 dar/30 ash/20 atr/2 ang	2+,3−
US730			2+,3−
US845	US388 × G695	33 cor/35 dar/8 ash/1 ang/25 ell/8 con	1+,2+,3+
US847			1+,2+,3+
US849	US388 × G478	35 cor/38 dar/2 ang/25 ell	3+,4+,5−

^{z}atr-*V. atrococcum*, dar-*V. darrowi*, ell-*V. elliottii*, myt-*V. myrtilloides*, ang-*V. angustifolium*, mys-*V. myrsinites*, ash-*V. ashei*.

characteristics appear highly heritable (Chandler et al., 1985; Erb et al., 1990 and 1991), temperature and mineral soil adaptations are only now being evaluated in the complex hybrids already released.

Several other wild species could potentially transmit mineral soil adaptations and drought tolerance including *V. angustifolium, V. ashei, V. pallidum, V. elliotii, V. myrtilloides* and *V. atrococcum* (Galletta, 1975; Korcak et al., 1982). Some of these have negative attributes such as low plant stature (*V. myrtilloides, V. angustifolium*), small fruit size (*V. elliottii, V. angustifolium*) and dark fruit (*V. atrococcum, V. myrtilloides*) (Ballington et al., 1984a), but these negative attributes can probably be eliminated through selective breeding. The extensive, complex germplasm developed by Draper affords many opportunities for recombination and enhancement.

Useful levels of disease and pest resistance may also be captured from wild species. Resistance to the sharp-nosed leaf hopper exist in *V. ashei* (Meyer and Ballington, 1990). Resistance to mummy berry and stem canker are found in wild *V. corymbosum* and *V. ashei* (Ballington, 1990). *V. corymbosum, V. ashei* and *V. elliotii* all carry resistance to stem blight (Buckley and Ballington, 1987). Draper was able to successfully use *V. angustifolium* and *V. atrococcum* as sources of resistance to *Phytophthora* root rot (Draper et al., 1971, 1972; Erb et al., 1987).

Genes for numerous other horticulturally important traits exist in the wild species (Luby et al., 1991). *V. elliottii, V. pallidum,* and *V. angustifolium* carry genes for early flowering and ripening. Some populations of *V. elliottii, V. constablaei,* and *V. ashei* have delayed bloom and are very late ripening (Ballington et al., 1984b and 1986). Extreme cold hardiness exists in *V. constablaei, V. angustifolium,* and *V. myrtilloides*. Most of these species are represented in the complex hybrids of Draper.

One of the major questions concerning the use of southern species is whether the complex hybrids will carry sufficient winter hardiness to be grown in northern climes. Whether or not southern species can be used in northern highbush breeding will depend on whether the genes regulating useful traits such as upland adaptation and heat tolerance can be teased apart from those associated with low chilling and sensitivity to cold. Draper has provided ample material to develop these recombinants by making a conscious effort to combine the genomes of both southern and northern species. For example, JU 60, 62 and 64 are hybrids of southern *V. myrsinites* from Florida and *V. angustifolium* from Maine. 'Sierra' is a blend of cold adapted *V. corymbosum* and *V. constablei* with southern *V. darrowi* and *V. ashei* (Figure 2).

Last year, we collected preliminary data that indicates Draper's complex hybrids do indeed contain a high degree of winter hardiness. In the spring of 1993, we established replicated plantings of Arlen's elite selections at Wooster, Ohio and East Lansing, MI. Four single plant replications were arranged in a randomized complete block design at each site, with plants spaced 1.2 × 3.0 m (4 × 10 ft) apart. Standard cultural practices were employed (Goulart, 1994). Winter temperatures gradually declined in the fall at both locations, but reached record lows of $-35°C$ ($-31°F$) at East Lansing and $-25°C$ ($-13°F$) at Wooster. Snow cover averaged in excess of 0.50 m (20 in) during most of the winter, but there were several relatively snow-free periods of extreme cold.

Clones were evaluated at bloom in 1994 for plant height, percentage of flower buds which produced developing flowers, and percentage of wood damaged by cold. Cultivars generally displayed greater wood than bud hardiness. The most hardy cultivars were those developed specifically for the north, either as highbush types (Patriot, Elliott) or half-highs (North Country, St. Cloud), while those developed for the south suffered the most damage (Cape Fear, Blueridge) (Table 2). However, some of the complex hybrids of Draper contained a considerable amount of winter hardiness. Most notably, 'Sierra' performed well at both locations, with as much bud and wood hardiness as the standard northern cultivar 'Jersey.' Most of the other complex hybrids were severely damaged in Michigan, but a few of them did well in Ohio (US 676, US 717, US 612). One of the most hardy, US 676, is composed of 50% southern species (Table 2).

We also screened winter hardiness in a 4-year-old breeding block at Benton Harbor, MI, containing adjacent rows of a F_2 population of US 75 selfed, a BC_1 population of US 75 × Bluecrop and 3 BC_2 populations composed of MSU-5, -6 and -7 × Bluecrop. The MSU selections had been selected earlier for their photosynthetic heat tolerance (Hancock et al., 1992). Winter conditions were similar to

East Lansing, except winter lows were slightly higher ($-30°C$ [$-22°F$]). Hardiness was calculated as the percentage of flower buds that developed normally in the spring.

A considerable amount of segregation for cold hardiness was observed, with some of the BC_2 individuals being completely hardy even though they were composed of 12.5% *V. darrowi* (Figure 3). Some of these hardy genotypes also had excellent fruit quality and yield. We have not screened these populations yet for their photosynthetic patterns, but they should provide a vehicle to test whether the cold tolerance of *V. corymbosum* can be genetically combined with the heat tolerance of *V. darrowi*.

It seems clear that Draper's complex hybrids will prove useful in the north, both as breeding parents and cultivars. His southern/ northern blend 'Sierra' appears adapted to severe winters and has excellent fruit quality. His other complex hybrids are not quite as hardy, but the breeding studies at MSU indicate that another round of crosses will probably solve this problem. One wonders how many more additional contributions will be made, as a broader array of Draper's inter-specific hybrids are carefully evaluated and utilized.

FIGURE 2. Pedigree of 'Sierra,' a complex Draper hybrid.

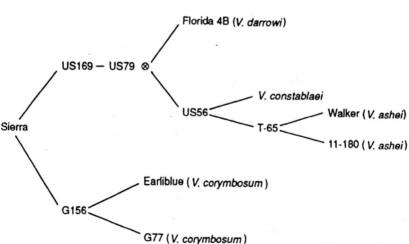

TABLE 2. Cold hardiness of cultivars and selected inter-specific hybrids of Draper.

Genotype	Winter Damage-Floral[x]		Winter Damage-Wood[y]		Plant Height[z]
	Ohio	Michigan	Ohio	Michigan	
Patriot	9.75	5.00	9.50	8.50	3
Elliott	9.25	4.30	9.50	7.30	4
North country	9.00	6.00	10.00	8.25	1
Sierra	9.00	3.00	9.75	8.30	2
Jersey	9.00	3.00	9.00	8.50	4
Bluecrop	8.25	1.80	9.50	7.70	4
St. Cloud	8.00	6.50	9.75	9.30	1
Mn 408	7.75	4.75	9.75	8.30	1
Spartan	7.50	3.50	8.75	8.00	4
Sunrise	7.00	1.30	9.50	6.00	1
Northland	6.75	4.00	9.50	7.50	3
US 676	6.75	1.00	7.75	3.50	3
US 717	6.50	1.30	7.75	5.25	2
US 612	6.30	1.30	8.75	3.30	4
US 702	4.75	1.25	7.25	2.30	2
Northblue	4.30	3.25	7.75	8.00	1
G 344	4.30	1.00	7.50	6.50	3
US 673	4.30	1.00	6.50	3.00	3
G 245	3.50	2.00	8.75	6.80	3
US 654	3.50	0.80	8.50	2.00	4
Bluetta	3.25	2.25	8.75	8.25	3
Blueridge	3.25	0.50	7.50	1.00	2
US 667	2.75	0.75	8.25	4.25	5
Cape Fear	2.75	0.80	8.00	3.75	2
US 671	2.50	1.30	6.25	5.80	3
US 665	2.00	1.00	7.75	2.00	5
US 720	1.30	0.80	1.75	1.00	3
US 730	1.00	0.80	2.75	1.80	3
US 729	1.00	0.80	1.25	2.50	3
US 723	1.00	0.50	1.25	0.50	4
LSD	2.14	1.35	1.50	2.35	

[x]1 = all buds destroyed, 10 = no buds destroyed.
[y]1 = dead to ground, 10 = no damage.
[z]1 = short (< 30 cm), 5 = tall (> 75 cm).

FIGURE 3. Hardiness of F_2, BC_1 and BC_2 populations of *V. darrowi* (Florida 4b) × *V. corymbosum* ('Bluecrop'). Winter hardiness ratings: 1 > 80% flower buds destroyed; 5 < 20% flower buds destroyed. MSU-5, 6 and 7 were the most heat tolerant individuals selected from the BC_1 population (Hancock et al., 1992).

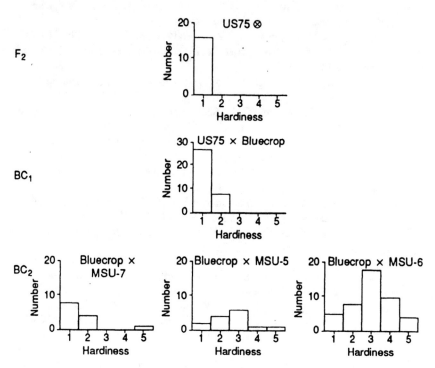

LITERATURE CITED

Ballington, J. M. 1990. Germplasm resources available to meet future needs for blueberry cultivar improvement. Fruit. Var. J. 44:54-63.

Ballington, J. R. and G. J. Galletta. 1978. Comparative crossability of diploid *Vaccinium* species. J. Amer. Soc. Hort. Sci. 103:544-560.

Ballington, J. R., Y. M. Isenberg, and A. D. Draper. 1986. Flowering and fruiting characteristics of *Vaccinium ashei* and *Vaccinium ashei-Vaccinium constablaei* derivative blueberry progenies. J. Amer. Soc. Hort. Sci. 111:950-955.

Ballington, J. R., W. E. Ballinger, W. H. Swallow, G. J. Galletta, and L. J. Kushman. 1984a. Fruit quality characterization of 11 *Vaccinium* species. J. Amer. Soc. Hort. Sci. 109:684-689.

Ballington, J. R., W. E. Ballinger, C. M. Mainland, W. H. Swallow, E. P. Maness,

G. J. Galletta, and L. J. Kushman. 1984b. Ripening season of *Vaccinium* species in southeastern North Carolina. J. Amer. Soc. Hort. Sci. 109:392-396.

Bruederle, L. P., N. Vorsa, and J. R. Ballington. 1991. Population genetic structure in diploid blueberry, *Vaccinium* section *Cyanococcus* (Ericaceae). Amer. J. Bot. 78:230-237.

Buckley, B. and J. R. Ballington. 1987. Screening native *Vaccinium* species for resistance to stem blight. HortScience 22:101.

Chandler, C. K., A. D. Draper, G. J. Galletta, and J. C. Bouwkamp. 1985. Combining ability of blueberry interspecific hybrids for growth on upland soils. HortScience 20:257-258.

Darrow, G. M. and W. H. Camp. 1945. *Vaccinium* hybrids and the development of horticultural material. Bull. Torrey Bot. Club 72:1-21.

Darrow, G. M., H. Derman, and D. H. Scott. 1949. A tetraploid blueberry from a cross of diploid and hexaploid species. J. Hered. 40:304-306.

Darrow, G. M., D. H. Scott, and H. Derman. 1954. Tetraploid blueberries from hexaploid × diploid species crosses. Proc. Amer. Soc. Hort. Sci. 63:266-270.

Draper, A. D. 1977. Tetraploid hybrids from crosses of diploid, tetraploid and hexaploid *Vaccinium* species. Acta Horticulturae 61:33-37.

Draper, A. D. and D. H. Scott. 1971. Inheritance of albino seedlings in tetraploid highbush blueberry. J. Amer. Soc. Hort. Sci. 96:791-792.

Draper, A. D., S. M. Mircetich, and D. H. Scott. 1971. *Vaccinium* clones resistant to *Phytophthora cinnamomi*. HortScience 6:167-169.

Draper, A. D., A. M. Stretch and D. H. Scott. 1972. Two tetraploid sources of resistance for breeding blueberries resistant to *Phytophthora cinnamomi* Rands. HortScience 7:266-268.

Draper, A. D., G. J. Galletta, and J. R. Ballington. 1982. Breeding methods for improving southern tetraploid blueberries. J. Amer. Soc. Hort. Sci. 107:106-109.

Eck, P. and N. F. Childers. 1966. Blueberry culture. Rutgers Univ. Press, New Brunswick, N.J.

Ehlenfeld, M. and N. Vorsa. 1993. Highbush = rabbitage = lowbush blueberries?!? ... Another perspective. Hort Technology 3:465-466.

Erb, W. A., J. N. Moore, and R. E. Sterne. 1987. Response of blueberry cultivars to inoculation with *Phytophthora cinnamomi* Rands Zoospores. HortScience 22:298-300.

Erb, W. A., A. D. Draper, and H. J. Swartz. 1988. Screening interspecific blueberry seedling populations for drought tolerance. J. Amer. Soc. Hort. Sci. 113:599-604.

Erb, W. A., A. D. Draper, G. J. Galletta, and H. J. Swartz. 1990. Combining ability for plant and fruit traits of interspecific blueberry progenies on mineral soil. J. Amer. Soc. Hort. Sci. 115:1025-1028.

Erb, W. A., A. D. Draper, and H. J. Swartz. 1991. Combining ability for canopy growth and gas exchange of interspecific blueberries under moderate water deficit. J. Amer. Soc. Hort. Sci. 116:564-573.

Erb, W. A., A. D. Draper, and H. J. Swartz. 1993. Relation between moisture stress and mineral soil tolerance in blueberries. J. Amer. Soc. Hort. Sci. 118:130-134.

Finn, C. E., J. J. Luby, and D. K. Wildung. 1990. Half-high blueberry cultivars. Fruit Varieties J. 44:63-68.

Galletta, G. J. 1975. Blueberries and cranberries, pp. 154-196. In: J. N. Moore and J. Janick (eds.) Advances in Fruit breeding. Purdue University Press, Lafayette, Ind.

Goldy, R. G. and P. M. Lyrene. 1984. Meiotic abnormalities of *Vaccinium ashei* × *Vaccinium darrowi* hybrids. Can. J. Bot. 26:146-151.

Goulart, B. L. 1994. Small fruit production and pest management guide. Penn State. College of Agricultural Sciences. University Park, Pennsylvania.

Hall, I. V. and L. E. Aalders. 1961. Cytotaxonomy of lowbush blueberries of eastern Canada. Amer. J. Bot. 48:199-201.

Hancock, J. F. 1993. The blueberry, *Vaccinium*. In: N. E. Simmonds (ed.). Evolution of crop plants. Longman, London (In press).

Hancock, J. F. and B. Goulart. 1993. A blueberry by any other name . . . HortTechnology 3:254-255.

Hancock, J. F., K. Haghighi, S. L. Krebs, J. A. Flore, and A. D. Draper. 1992. Photosynthetic heat stability in highbush blueberries and the possibility of genetic improvement. HortScience 27:1111-1113.

Hanson, E. A. and J. F. Hancock. 1990. Highbush blueberry cultivars and production trends. Fruit Var. J. 44:77-81.

Jelenkovic, G. and A. D. Draper. 1973. Breeding value of pentaploid interspecific hybrids of *Vaccinium*, pp. 237-244. In: Proc. Intern. Symposium on Breeding *Ribes, Rubus,* and *Vaccinium*. Yugoslavian Society for Horticultural Science.

Korcak, R. F. 1986a. Adaptability of blueberry species to various soil types: I. Growth and initial fruiting. J. Amer. Soc. Hort. Sci. 111:816-821.

Korcak, R. F. 1986b. Adaptability of blueberry species to various soil types: II. Leaf and soil analysis. J. Amer. Soc. Hort. Sci. 111:822-828.

Korcak, R. F. 1989a. Variation in the nutrition of blueberries and other calcifuges. HortScience 24:573-578.

Korcak, R. F. 1989b. Influence of micronutrient and phosphorous levels and chelator to iron ratio on growth, chlorosis and nutrition of *Vaccinium ashei,* Reade and *V. elliottii,* Chapman. J. Plant Nutrition 12:1311-1320.

Korcak, R. F., G. J. Galletta, and A. D. Draper. 1982. Response of blueberry seedlings to a range of soil types. J. Amer. Soc. Hort. Sci. 107:1153-1160.

Krebs, S. L. and J. F. Hancock. 1989. Tetrasomic inheritance of isoenzyme markers in the highbush blueberry, *Vaccinium corymbosum* L. Heredity 11-18.

Luby, J. J., J. R. Ballington, A. D. Draper, K. Pliszka, and M. E. Austin. 1991. Blueberries and cranberries (*Vaccinium*). *In:* J. N. Moore and J. R. Ballington, Jr. (eds.). Genetic Resources of Temperate Fruit and Nut Crops. Int. Soc. Hort. Sci., Wageningen.

Lyrene, P. M. 1990. Low-chill highbush blueberries. Fruit Varieties J. 44:82-86.

Lyrene, P. M. and W. B. Sherman. 1983. Mitotic instability and 2n gamete production in *Vaccinium corymbosum* × *V. elliottii* hybrids. J. Amer. Soc. Hort. Sci. 108:339-342.

Lyrene, P. M. and J. R. Ballington. 1986. Wide hybridization in *Vaccinium*. HortScience 21:52-57.

Mengalos, B. S. and J. R. Ballington. 1988. Unreduced pollen frequencies vs. hybrid production in diploid-tetraploid *Vaccinium* crosses. Euphytica 39: 271-278.

Meyer, J. R and J. R. Ballington. 1990. Resistance of *Vaccinium* spp. to the leafhopper *Scaphytopius magdalensis* (Homoptera: Cicadellidae). Annals Ent. Society America 83:515-520.

Moon, J. W., J. A. Flore, and J. F. Hancock. 1987a. A comparison of carbon and water vapor gas exchange characteristics between a diploid lowbush blueberry and highbush blueberry. J. Amer. Soc. Hort. Sci. 112:134-139.

Moon, J. W., J. F. Hancock, A. D. Draper, and J. A. Flore. 1987b. Genotypic differences in the effect of temperature on CO_2 assimilation and water use efficiency in blueberry. J. Amer. Soc. Hort. Sci. 112:170-173.

Moore, J. N. 1975. Improving highbush blueberries by breeding and selection. Euphytica 14:39-48.

Ortiz, R., N. Vorsa, L. P. Bruederle, and T. Laverty. 1992. Occurrence of unreduced pollen in diploid blueberry species, *Vaccinium* section *Cyanococcus*. Theo. Appl. Genet. 85:55-60.

Sharp, R. H. and W. B. Sherman. 1971. Breeding blueberries for low chilling requirement. HortScience 6:145-147.

Vander Kloet, S. P. 1980. The taxonomy of the highbush blueberry, *Vaccinium corymbosum*. Can. J. Bot. 58:1187-1201.

Vander Kloet, S. P. 1988. The genus *Vaccinium* in North America. Agriculture Canada Publ. 1828. 201 pp.

Vorsa, N. 1987. Meiotic chromosomal pairing and irregularities in blueberry interspecific backcross-1 hybrids. J. Heredity 78:395-399.

Vorsa, N. 1988. Differential transmission of extra genomic chromosomes in pentaploid blueberry. Theo. Appl. Genet. 75:585-591.

Vorsa, N., G. Jelenkovic, A. D. Draper, and M. V. Welker. 1987. Fertility of 4x − 5x and 5x − 4x progenies derived from *Vaccinium ashei/corymbosum* pentaploid hybrids. J. Amer. Soc. Hort. Sci. 112:993-997.

Vorsa, N., P. S. Manos, and M. I. van Heemstra. 1988. Isozyme variation and inheritance in blueberry. Genome 30:776.

In Search of the Perfect Blueberry Variety

Arlen D. Draper

The quest for the perfect blueberry variety for me started 30 years ago when I was still filled with the wisdom of graduate school and the brashness of youth. It has led me through, not only the valleys of spring frosts, droughts, and winter injury, but also to the peaks of some fine germplasm, some serendipitous crosses, and some treasured friendships. I've had the good fortune to have worked with some very competent and good people. I am especially grateful to Dr. Donald Scott who patiently labored at converting me from a forage breeder to a small fruits breeder. Dr. George Darrow's desk was in my office and I looked forward to his weekly visits so I could pump him for information. It was a privilege to work with the staff of the Atlantic Blueberry Company. They shouldered much of the responsibility for seedling care and selection propagation for cultivar releases, and tried hard to keep me focused on the qualities of a good blueberry as seen through the eyes of a commercial grower.

Early in my blueberry work it appeared that only small gains were to be made by continuing crossing and selecting within cultivated *Vaccinium corymbosum*. That pot had been stirred long enough. We obtained selections of a number of native *Vaccinium* species that had characteristics of commercial interest. For example we made crosses using *V. angustifolium* and *V. myrtilloides* and

Arlen D. Draper is a retired plant breeder, USDA-ARS. He presently resides at 604 East Park Drive, Payson, AZ 85541.

[Haworth co-indexing entry note]: "In Search of the Perfect Blueberry Variety." Draper, Arlen D. Co-published simultaneously in *Journal of Small Fruit & Viticulture* (Food Products Press, an imprint of The Haworth Press, Inc.) Vol. 3, No. 2/3, 1995, pp. 17-20; and: *Blueberries: A Century of Research* (ed: Robert E. Gough, and Ronald F. Korcak) Food Products Press, an imprint of The Haworth Press, Inc., 1995, pp. 17-20. Single or multiple copies of this article are available from The Haworth Document Delivery Service [1-800-342-9678, 9:00 a.m. - 5:00 p.m. (EST)].

native *V. corymbosum* from Maine and Pennsylvania, looking for greater winter hardiness for northern blueberries. For southern highbush blueberries we needed low chilling requirement, disease resistance, increased soil adaptation, and heat tolerance. For this we looked to the native southern species *V. darrowi, V. myrsinites, V. elliottii,* and *V. ashei.* We probed *V. atrococcum* for early ripening, *V. constablaei* for short interval from bloom to ripe fruit and *V. simulatum* for cold hardiness.

Though these species range from the diploid to the hexaploid chromosome numbers, our objective was to end with tetraploid hybrids that could be crossed with *V. corymbosum* to take advantage of its 50-year improvement in fruit quality. To obtain tetraploid germplasm we used the following procedures: chromosome doubling of diploids with colchicine, crossing diploids × tetraploids looking for unreduced gamete production in the diploid, and making diploid × hexaploid crosses hoping for tetraploid seedlings.

The most successful technique proved to be the crossing of diploid × tetraploid genotypes. A Florida selection of *V. darrowi*, 4b, crossed with 'Bluecrop' produced tetraploid hybrids that in one backcross generation provided selections with commercial qualities for the southern areas.

Two unexpected developments occurred in the crosses of species of differing ploidy levels–diploid seedlings were produced in diploids × tetraploids and in diploid × hexaploid crosses. Another unexpected development occurred in the diploid × hexaploid crosses, all hybrids proved to be pentaploid hybrids except for the rare diploid.

Some other results were: *V. atrococcum* produced seedlings that were early ripening, but the fruit was soft and dark. The F_2 populations of *V. angustifolium* × *V. corymbosum* did not segregate for plant height, all were of short stature. The seedlings were productive, but were too short. The same was true of the *V. myrsinites* hybrid seedlings. They were productive and well adapted to southern environments, but were too short to be machine harvested. We learned that a species difficult to cross with *V. corymbosum*, such as *V. elliottii,* could be crossed with 'Florida 4b,' and the F_1 hybrids could then be crossed to *V. corymbosum. V. constablaei* has contributed earlier ripening without causing earlier flowering to the *V.*

ashei germplasm and that should soon be seen in commercial cultivars. The extremely winterhardy native *V. corymbosum* selections
from Maine and Pennsylvania proved not to be winterhardy in
mid-Atlantic areas with fluctuating late-winter temperatures. These
populations gave us insight, and what we had considered to be a
lack of winterhardiness was in reality bud activity starting too
quickly during brief warm spells followed by cold temperatures in
late winter or early spring.

From the interspecific hybridization there came an unexpected
increase in fruit quality (scar, firmness, good flavor retention),
along with wider soil and climatic adaptation and increased productivity. One regret is that we didn't try harder to incorporate *V.
ashei* germplasm into the southern highbush. It was the current
dogma that the F_1 hybrids were sterile and a deadend. Many of the
progenies of this cross were sterile, but certain crosses produced
fully fertile hybrids. These were usually vigorous plants, with
good fruit and small scars, but on the dark side. One of the selections at the Poplarville Small Fruits Research Station is an F_1
hybrid and is slated to be released soon. Another regret is that I
was unsuccessful in using *V. membraneceum* and *V. deliciosum* in
the interspecific crossing because the plants refused to grow in
Beltsville.

In the search for the perfect blueberry variety we sorted through
hundreds of thousands of seedlings grown from thousands of crosses producing thousands of selections. It has truthfully been said that
you must kiss a lot of frogs in order to find the prince. There were
moments of inspiration along the way when I was prepared to
receive them, usually coming during sleepless nights or early in the
morning. Some of the perfect crosses that I dreamed up on my own
usually proved to be something less than perfect. I have often been
frustrated by failing to find the perfect variety. But now I am beginning to realize that it was the search that was important, more so
than the finding. I thrilled with the hope when walking seedling
rows that a world-beater was just a few plants ahead or perhaps in
the next row. This is not at all unlike the old prospector looking for
gold and always hoping to make that big strike just over the next
hill. In reality, I probably walked past several fine varieties not
knowing how to recognize them. All in all, it has been a wonderful

journey. I want to express my appreciation to all with whom I worked or associated, for your kindness and for the things that you taught me. I didn't find the perfect variety, but I hope that the ones that I had a part in developing will find a useful niche.

The Deerberry
[*Vaccinium stamineum* L. *Vaccinium* Section *Polycodium* (Raf.) Sleumer]: A Potential New Small Fruit Crop

James R. Ballington

SUMMARY. *Vaccinium stamineum* L. is an extremely polymorphic diploid species adapted to acidic dry uplands in eastern North America. It is insect pollinated and has unique floral morphology including open aestivation of flowers in the bud. Plant habit varies from stoloniferous to essentially monostem. Fruit size is large, fruit color and glaucousness extremely variable and the fruit usually falls off the plant (shatters) at maturity. The skin of the fruit typically has a bitter taste although the degree of bitterness varies within and among localities. Flavor, anthocyanins and flavonols of fruits are very similar to cranberry. Unlike cranberry, soluble solids content can be quite high. Elite genotypes for large fruit size, glaucousness, palatability and resistance to fruit shattering have been identified in Lexington County, SC. In the southern US the fruits have been used locally for jams, jellies and pies for many years. It appears that *V. stamineum* has potential as a new commercial small fruit crop, especially for use in processed products. *[Article copies available from The Haworth Document Delivery Service: 1-800-342-9678.]*

KEYWORDS: Gooseberry; Cranberry; Blueberry; *V. myrtillus*; *V. macrocarpon*

James R. Ballington, Horticultural Science Department, North Carolina State University, Raleigh, NC 27695-7609.

[Haworth co-indexing entry note]: "The Deerberry [*Vaccinium stamineum* L. *Vaccinium* Section *Polycodium* (Raf.) Sleumer]: A Potential New Small Fruit Crop." Ballington, James R. Co-published simultaneously in *Journal of Small Fruit & Viticulture* (Food Products Press, an imprint of The Haworth Press, Inc.) Vol. 3, No. 2/3, 1995, pp. 21-28; and: *Blueberries: A Century of Research* (ed: Robert E. Gough, and Ronald F. Korcak) Food Products Press, an imprint of The Haworth Press, Inc., 1995, pp. 21-28. Single or multiple copies of this article are available from The Haworth Document Delivery Service [1-800-342-9678, 9:00 a.m. - 5:00 p.m. (EST)].

INTRODUCTION

When Europeans began building permanent settlements in eastern North America they encountered a wide array of indigenous small fruit species. A number of these species, especially blackberries and red raspberries *(Rubus* spp.), were fairly similar to the ones they had known in Europe. Currants and gooseberries *(Ribes* spp.) were also fairly common in more northern areas and at higher elevations in the southern Appalachian mountains. However, *Vaccinium* species were both distinct from and much more diverse than anything that occurred in Europe. These included the large cranberry *(V. macrocarpon* Ait.), the various species of blueberries, and on dry acidic uplands, a species which in most regions is commonly referred to as the "deerberry"(*V. stamineum* L.). In what is now the southeastern US, true gooseberries *(Ribes* spp.) were very rare, and early European settlers there considered the fruits of *V. stamineum* to be similar to gooseberries and began calling them by that name. *Vaccinium stamineum* is still referred to as "gooseberry" in the southern US, and fruit has been harvested from the wild for many years for use in jams, jellies, and also for making "gooseberry pies." The population density of *V. stamineum* is quite high in some areas, especially in the southern US. In the sandhill region of central South Carolina plants occur in such abundance that fruit was harvested for the commercial production of illegal "gooseberry brandy" during Prohibition. The purpose of the present paper is to describe the characteristics of *V. stamineum* and present the rationale for its potential as a new small fruit crop.

TAXONOMY AND DISTRIBUTION

Species designation in deerberries has been controversial, as has often been the case with other groups in *Vaccinium*. At one extreme Ashe (1931) recognized 21 species, while recent treatments (Baker, 1970; Vander Kloet, 1988) concluded that the entire group consisted of a single extremely polymorphic species, *V. stamineum*, which made up the monotypic section *Polycodium* (Raf.) Sleumer of *Vaccinium* L. All chromosome counts to date indicate that the species is diploid with $2n = 2x = 24$ chromosomes (Baker, 1970).

A number of characteristics distinguish *V. stamineum* from other groups of *Vaccinium* in eastern North America. One of these is dimorphic branches in which lateral and subterminal branches are either vegetative or reproductive (Baker, 1970). The reproductive branches are racemes with flowers occurring singly in the axils of persistent leafy bracts. Flowers are campanulate in shape and each emerges from the axil of a leafy bract as an already open, but miniature, version of the full size flower. Open aestivation of flowers is unique to this section in *Vaccinium* (Baker, 1970). In mature flowers, anther tubes project beyond the corolla as much as 3 mm (hence the specific epithet "stamineum") and the style projects 2 to 3 mm beyond the anther tubes (Baker, 1970). Flowers are insect-pollinated, in particular by bees that collect pollen by sonication of anthers (buzz-pollination) (Cane et al., 1985). In the present author's experience, the species is basically self-unfruitful.

Vaccinium stamineum occurs on dry uplands from New England and southern Ontario southward to Florida and Texas (Baker, 1970), and disjunct populations also occur in central Mexico (Vander Kloet, 1988). Although Baker (1970) could not find sufficient discontinuities to identify distinct populations morphologically, he did identify a number of "centers" in the southeastern US where specific gene and morphological combinations prevailed. In every case he found overlap among adjoining "centers." This, in conjunction with examination of herbarium specimens, lead him to conclude that the species basically forms a continuous population throughout its range.

PLANT AND FRUIT CHARACTERISTICS

Plant habit is extremely variable, from stoloniferous shrubs less than 0.5 m tall, to clump-forming suckering forms from 1.0 to 2.0 m tall to crown-forming and sometimes essentially monostem forms to as much as 4.0 m tall (Baker, 1970). It is not uncommon for the stoloniferous, clump-forming and crown-forming habit to be found among plants in a single sample collected within an area of 1000 m². Stoloniferous and suckering forms respond to fire and other disturbances such as logging and mowing similarly to stoloniferous *Vaccinium* sections *Cyanococcus* and *Myrtillus* species. Baker (1970) found that frequent disturbances such as mowing increased the percentage of reproductive branches. He observed shrubs only 2 dm tall that were composed almost entirely of flowering axes. Sharpe and Sherman (1971) also observed that plants of *V. stamineum* are capable of maturing full crops of fruit even when growing in deep droughty sands.

Fruit size is quite large for a wild *Vaccinium*, especially a diploid species. The general range in fruit diameters is from 5 mm to 16 mm, with occasional plants with fruit 19 mm in diameter (Baker, 1970; Ballington et al., 1984). Fruit size is positively correlated with number of developing seeds, and potentially viable seed numbers range from 5 to 29 per berry (Baker, 1970). Darrow (1941) found that seed size was approximately the same as *V. ashei* Reade. In the southeastern US, Baker (1970) determined that plants with the largest fruits on average occurred in the Lexington County, South Carolina and the coastal North Carolina/South Carolina border area. He also observed that fruit shape ranged from oblate to globose to ovoid to pyriform.

The color of mature fruit is extremely variable, ranging from greenish-white, yellowish, light red, dark red, blue, purple-blue, purple-black, reddish-black, to dark purple (Baker, 1970; Ballinger et al., 1981). The surface of the fruit may be either glaucous or glabrous, and glaucousness on the fruit is generally correlated with glaucousness on vegetative parts. Fruits from plants in the South Carolina sandhill region were the most glaucous observed by Baker (1970).

Fruits of *V. stamineum* typically have some bitterness in the skin, although the degree of bitterness and sweetness is quite variable. Baker (1970) determined that fruits from the South Carolina sandhill region were most palatable. Within this same region I have identified purple-fruited individuals with only slight bitterness and white-fruited forms completely free of bitterness with a very pleasant "peary" flavor. The flavor of the best purple-fruited forms resembles *V. myrtillus* L.

· Fruits typically drop to the ground as soon as they ripen and since the calyx tube is continuous with the pedicel, the pedicel remains attached to the fruit (Vander Kloet, 1988). Ballington et al. (1984)

identified non-shattering genotypes which will retain the fruit on the plant until all are ripe and can be harvested in a single operation.

Steyermark (1963) noted that the fruit of *V. stamineum* served cold after being cooked, has a flavor suggesting cranberry and gooseberry sauce combined with grapefruit marmalade. Ballinger et al. (1981) also noted that the fruit resembles cranberries in fresh market and culinary quality as well as anthocyanin and flavonol content. In contrast to cranberry, the soluble solids content of *V. stamineum* can be quite high (Ballington et al., 1984). Occasional genotypes from Lexington County, SC, have pigmented flesh like *V. myrtillus* (Luby et al., 1990).

DIFFICULTIES AND UNKNOWNS

As with any potential candidate for domestication, there are always large gaps in the information needed to more fully evaluate its feasibility and requirements, as well as difficulties to overcome. A number of these as regards *V. stamineum* are briefly discussed as follows:

- I have been growing deerberries in cultivation for over 15 years, but always as a minor component in blueberry selection trials. The *V. stamineum* plants in each trial have generally responded favorably to the cultural conditions and plant nutrient levels applied to the blueberry plants; however, specific nutrient needs and requirements are unknown. Likewise, no information is available on pruning requirements other than Baker's (1970) comment that regular mowing of low-growing forms significantly increased the percentage of reproductive axes.
- *Vaccinium stamineum* is cross-pollinated, and the possible existence of cross-incompatibility alleles is unknown. This could be important in selecting the genotypes to be established in a planting. Cane et al. (1985) indicated that a specific oligolectic bee is primarily responsible for pollination of deerberry. If wide scale commercial planting of *V. stamineum* were to occur, would the primary pollinator be available in sufficient numbers for adequate fruit set? A second question is, could secondary or more generalist pollinators adequately replace the primary pollinator if its numbers were insufficient?

- Information on diseases affecting *V. stamineum* is essentially non-existent. Genotypes from south Georgia and north Florida were infected by the blueberry stunt mycoplasma at Castle Hayne, NC, in the mid 1970s. Insect information is also very limited. Meyer and Ballington (1990) found that several genotypes from Lexington County, SC, showed high levels of non-feeding preference resistance to the sharp-nosed leafhopper vector of blueberry stunt. *Vaccinium stamineum* also has a specific maggot fly species or subspecies *(Rhagoletis* sp.) whose larva infect the fruit. Baker (1970) reported that a species of yellow necked caterpillar *(Datana* sp.) completely defoliated *V. stamineum* plants on several sites.
- While collected fruit of *V. stamineum* has been utilized locally in various home processed products for years, it has not been available in the quantities needed for controlled postharvest handling and processed product studies.
- Suckering and stoloniferous forms of *V. stamineum* can be clonally propagated readily by rhizome or sucker pieces collected in late winter. However, only a limited number of propagules are available from most mother plants. Propagation by leafy stem cuttings has been difficult with variable and generally low percentages of success in my experience. Another difficulty is that all clonal propagules have been extremely precocious, making development of mature size plants slow. Micropropagation may possibly be the solution to large-scale clonal propagation, if it can be shown to be economically feasible.
- Finally, it will be essential to incorporate the non-shattering fruit trait into any genotypes destined to be commercial cultivars. This has been an essential step in the domestication of many of the world's most important crop species. As noted earlier, sources of genes for non-shattering fruit have been identified in *V. stamineum.*

POTENTIAL UTILITY

Desirable fruit characteristics and a wide range of plant morphologies, along with upland adaptation and drought tolerance, indicate that *V. stamineum* has potential for development as a new small fruit crop. Its greatest potential initially would appear to be primarily in processed products. Since it has the same anthocyanin and flavonol content as cranberry, it should be the ideal fruit for blending with the latter in cranberry products. The high soluble solids content of *V. stamineum* fruit could also aid in reducing the sugar necessary to add to cranberry or other fruit products.

Two systems would appear to be realistic possibilities for managing *V. stamineum* as a commercial fruit crop for processing. Both of these assume that the non-shattering fruit trait has been incorporated into all genotypes involved. The first possibility involves establishment of narrow crown-forming upright growing genotypes in rows similar to current commercial blueberry plantings for once-over or multiple harvesting with over-the-row blueberry harvesters. The second possibility involves establishing uniform ripening low stoloniferous forms in matted-row type culture for once-over harvesting with dry cranberry or lowbush blueberry mechanical harvesters.

LITERATURE CITED

Ashe, W. W. 1931. *Polycodium*. J. Elisha Mitchell Scient. Soc. 46:196-213.

Baker, P. C. 1970. A systematic study of the genus *Vaccinium* sect. *Polycodium* (Raf.) Sleumer, in the southeastern United States. PhD Dissertation. University of North Carolina, Chapel Hill, NC. 248 pp.

Ballinger, W. E., E. P. Maness, and J. R. Ballington. 1981. Anthocyanin and total flavonol content of *Vaccinium stamineum* L. fruit. Sci. Hortic. 15:173-178.

Ballington, J. R., W. E. Ballinger, W. H. Swallow, G. J. Galletta, and L. J. Kushman. 1984. Fruit quality characterization of 11 *Vaccinium* species. J. Amer. Soc. Hort. Sci. 109:684-689.

Cane, J. H., G. C. Eichwort, F. R. Wesley and J. Spielholz. 1985. Pollination ecology of *Vaccinium stamineum* (Ericaceae: Vaccinioideae). Am. J. Bot. 72:135-142.

Darrow, G. M. 1941. Seed size in blueberry and related species. Proc. Amer. Soc. Hort. Sci. 38:438-440.

Luby, J. J., J. R. Ballington, A. D. Draper, K. Pliszka, and M. E. Austin. 1990. Blueberries and cranberries *(Vaccinium)*. pp. 393-456. *In:* J. N. Moore and J. R. Ballington, eds. Genetic Resources of Temperate Fruit and Nut Crops. ISHS, Wageningen, The Netherlands.

Meyer, J. R. and J. R. Ballington. 1990. Resistance of *Vaccinium* spp. to the

leafhopper *Scaphytopius magdalensis* (Homoptera: Cicadellidae). Ann. Entomol. Soc. Am. 83:515-520.

Sharpe, R. H. and W. B. Sherman. 1971. Breeding blueberries for low chilling
requirement. HortScience 6:145-147.

Steyermark, J. A. 1963. Flora of Missouri. Iowa State University Press, Ames, IA.

Vander Kloet, S. P. 1988. The genus *Vaccinium* in North America. Agriculture
Canada Publ. 1828. 201 pp.

SESSION II:
BLUEBERRY GENETICS
Moderator: Rebecca Darnell

Use of Sparkleberry
in Breeding Highbush Blueberry Cultivars

Paul M. Lyrene
Sylvia J. Brooks

SUMMARY. Diploid hybrids between *V. darrowi* Camp and *V. arboreum* Marsh. (sparkleberry) were allowed to open-pollinate in the presence of a wide range of blueberry cultivars and species at the Horticultural Research Unit in Gainesville, Florida. Open-pollinated seedlings (MIKs) with high fertility were selected and crossed to tetraploid southern highbush cultivars. Most crosses of this type gave numerous seedlings, indicating that the fertile MIKs were tetraploid

Paul M. Lyrene is Professor, and Sylvia J. Brooks is a graduate research assistant, Horticultural Sciences Department, Florida Agricultural Experiment Station, Gainesville, FL 32611.
Journal Series No. R-03952.

[Haworth co-indexing entry note]: "Use of Sparkleberry in Breeding Highbush Blueberry Cultivars." Lyrene, Paul M., and Sylvia J. Brooks. Co-published simultaneously in *Journal of Small Fruit & Viticulture* (Food Products Press, an imprint of The Haworth Press, Inc.) Vol. 3, No. 2/3, 1995, pp. 29-38; and: *Blueberries: A Century of Research* (ed: Robert E. Gough, and Ronald F. Korcak) Food Products Press, an imprint of The Haworth Press, Inc., 1995, pp. 29-38. Single or multiple copies of this article are available from The Haworth Document Delivery Service [1-800-342-9678, 9:00 a.m. - 5:00 p.m. (EST)].

29

or near tetraploid. The vigor of these backcross seedlings varied widely, both within and among crosses of MIK × highbush cultivar. The best seedlings from the best crosses had fruit of cultivar quality. Some seedlings ripened as early as the earliest southern highbush cultivars; others ripened later than the latest southern highbush cultivars. This study suggests that commercial cultivars could be selected in which one-eighth or more of the pedigree consists of *V. arboreum,* but many seedlings will have to be grown to sample all the plant types that can be derived from this gene pool. The extent to which these seedlings have inherited the superior tolerance of *V. arboreum* to droughty, low-organic, high pH soils is unknown. *[Article copies available from The Haworth Document Delivery Service: 1-800-342-9678.]*

KEYWORDS: Blueberry breeding; Blueberry genetics; Interspecific hybrids; Sparkleberry

INTRODUCTION

Highbush blueberry cultivars are based largely on tetraploid *V. corymbosum* selections from areas with acid, moist, organic soils in southern New Jersey (Coville, 1937). These selections were from a race of *V. corymbosum* that was given species status as *V. australe* Small by Camp (1945), but *V. australe* is now considered by many to be a race of *V. corymbosum* (Luby et al., 1990).

Southern highbush cultivars were developed by introgressing genes from low-chill blueberry selections, primarily *V. darrowi* Camp from Florida, into the cultivated northern highbush gene pool developed in New Jersey and Michigan (Sharpe, 1954; Sharpe and Darrow, 1959; Sharpe and Sherman, 1971). Use of *V. darrowi* in breeding highbush blueberries has given selections with excellent berry quality, zero to low chill requirement, a tendency to leaf luxuriously in the spring and the ability to continue growth through the heat of the north-Florida summer. As a source of drought tolerance or tolerance to soils high in pH or low in organic-matter, *V. darrowi* has been less useful. The apparent drought tolerance of *V. darrowi* in its Florida native habitat is probably a consequence of its small leaves and the high percentage of its biomass that consists of roots and rhizomes. A large collection of *V. darrowi* clones propagated by cuttings and maintained at the Horticultural Unit in Gainesville shows no more foliage drought tolerance than *V. corymb-*

osum collected from moist, organic soils in north Florida and southeastern Georgia. However, well-established *V. darrowi* would probably resprout from underground if above-ground stems were killed by drought, whereas *V. corymbosum* would probably not.

Although useful in breeding southern highbush cultivars, *V. darrowi* has contributed some undesirable features to its hybrids which have required much selection to mitigate. These include low growth habit and a tendency to overproduce to the extent that the plants die or produce small, low-quality fruit unless much of the crop is removed by freezes or severe winter pruning. Tolerance to phytophthora root rot (*Phytophthora cinnamomi* Rands) and stem blight (*Botryosphaeria dothedia* (Moug. ex Fr.) Ces. and deNot.) have also been erratic in *V. darrowi*-derived southern highbush.

In 1981, experiments were initiated to test the possibility of using sparkleberry and deerberry (*V. stamineum* L.) in breeding highbush blueberry cultivars tolerant to warm soils low in organic matter. Both species were crossed with *Vaccinium* section Cyanococcus species *V. elliottii* and *V. darrowi*. These Cyanococcus species were selected because their chromosome numbers (diploid) matched that of deerberry and sparkleberry, a factor that was expected to favor hybridization. All four combinations gave hybrids, but hybrids with *V. stamineum* showed little vigor, whereas some hybrids with *V. arboreum* were highly vigorous and persistent in the field and flowered heavily each year (Lyrene, 1991). The *V. elliottii* × *V. arboreum* hybrids gave no viable seed when open-pollinated in the field, so further work was concentrated on the *V. darrowi* × *V. arboreum* hybrids. This report describes progress that has been made in introgressing genes from *V. darrowi* × *V. arboreum* hybrids into the cultivated southern highbush gene pool.

MATERIALS AND METHODS

Two *V. darrowi* clones and a diploid hybrid of *V. darrowi* × diploid Florida *V. corymbosum* (*V. fuscatum* Ait. race) were pollinated in a greenhouse with pollen from 6 Florida sparkleberry plants (Lyrene, 1991). The resulting seedlings were grown to the age of flowering in a high-density field nursery (Sherman et al.,

1973). Sixteen of the most vigorous hybrids were selected and transplanted to a 1.5 m × 4 m spacing in December 1984. Hybrid characteristics included low fertility and intermediacy in bark, leaf, flower, inflorescence and fruit phenotypes and in timing of flowering, fruiting, leafing and defoliation, as well as intermediacy in plant size and shape. Over the next decade, many of the plants flowered heavily every year, and, in years without late freezes, they produced an appreciable number of berries after open-pollination. Seeds were extracted from these berries, and the seedlings were grown and evaluated in high-density nurseries. More than 100 seedlings, termed MIKs (Mother Is Known), were raised to flowering age from seed collected in 4 different years. The MIKs were evaluated for vigor, fruit quality, and seed production in high-density field nurseries that also contained numerous diploid, tetraploid, and hexaploid blueberry pollen sources. The most vigorous and fruitful MIKs that had the largest, highest-quality fruit were selected for crossing to southern highbush cultivars and breeding lines. Over 1000 seedlings from 16 crosses were planted in high-density field nurseries, and over half survived to flowering age. Some of the most fertile MIKs were also self-pollinated and intercrossed, and seedlings were evaluated in high-density nurseries. A number of MIKs were propagated by cuttings and several were transplanted to field nurseries at a 1.5 × 4 m spacing.

RESULTS AND DISCUSSION

Characteristics of Sparkleberry

The sparkleberry is a shrub or small tree native in the southeastern United States from Florida to Indiana and through the Ozarks to the Edwards Plateau, Texas (Camp, 1945). In Florida, the plants are often monopodial and up to 10 m tall, but on land subject to fire, they may form colonies with many stalks. In native Florida woodlands, sparkleberries are commonly found growing with highbush blueberries, rabbiteye blueberries, and *V. elliottii*, but they also occur where higher soil pH or excessive soil drainage exclude the other three. On droughty soils in north Florida they frequently co-occur with deerberries (*V. stamineum* L.). In Texas, Stockton (1976) found that sparkleberries often grew on soils where the pH was too high for other blueberry species. Sparkleberries were found growing without chlorosis on soils where the pH was as high as 6.4. Superior root characteristics have led to interest in the use of sparkleberry as a rootstock onto which highbush and rabbiteye cultivars can be grafted (Ballington, 1990).

Compared to the seeds of *Vaccinium* species in section Cyanococcus, sparkleberry seeds are hard to germinate. We have had fair to good results after soaking the seeds overnight in gibberellic acid solution (4 grams per liter of water) and then planting under conditions favorable for seed germination in Cyanococcus species (placed uncovered on sphagnum peat in an unheated greenhouse in November in Gainesville, FL, and watered with intermittent mist from 1200 to 1500 each day for 2 months). Compared to Cyanococcus seedlings, sparkleberry seedlings grow poorly in pots of pure sphagnum peat. However, when transplanted to sandy soil in a field nursery, they overtake and surpass most Cyanococcus seedlings within two years.

Sparkleberries are harder to propagate from stem cuttings than Cyanococcus species. Some sparkleberry ramets we propagated from softwood stem cuttings of mature plants showed extreme plagiotropy, far more than we have seen with Cyanococcus species. At least one MIK shows strong plagiotropy when propagated by softwood cuttings taken from lateral branches of the ortet. In contrast to cuttings from fruiting sparkleberry trees, juvenile softwood cuttings taken from sparkleberry seedlings less than 6 months old rooted readily under mist in our greenhouse.

Sparkleberry seedlings at the Horticultural Research Unit in Gainesville have had excellent growth and survival compared to highbush seedlings. The sparkleberry seedlings produce deep, large-diameter roots, in contrast with the shallow, netted root system of highbush seedlings. Despite the fact that a full-grown sparkleberry is more a tree than a shrub, sparkleberry seedlings are quite precocious in flowering and fruiting. Most sparkleberry seedlings flower a year-and-a-half from seed sown in Florida during November, whether grown in pots or in high-density field nurseries.

Characteristics of F1 Hybrids

The F1 *V. darrowi* × *V. arboreum* hybrids have been vigorous and persistent in the field for over 10 years, and most are over 3 m tall. In years when winter temperatures are above normal in Gainesville, they begin flowering in early January and continue flowering through mid-April. Peak flowering normally occurs in March. Some plants flower abundantly, but all have low percent fruit set. Some plants produce no fruit, and even on the most fruitful, fewer than 10% of the flowers produce ripe fruit. Still, after heavy flowering, over 1000 berries per plant per year have been harvested from the most fruitful plants in several years that had no freezes in February or March. Most of the berries have one or no well-developed seeds. Several sources of pollen have been available to produce these open-pollinated seeds. Numerous diploid, tetraploid, and hexaploid blueberry species and selections flower nearby during the first half of the flowering season of the *V. darrowi* × *V. arboreum* hybrids. During the second half of the flowering period for the hybrids, the pollen supply from other blueberries is lower and is mainly from highchill highbush.

Characteristics of MIKs

Open-pollinated progeny from the *V. darrowi* × *V. arboreum* F1 hybrids are highly variable. Fruitfulness after open-pollination in the field ranges from zero to nearly 100% for different seedlings. About half the seedlings produce little or no fruit, and fewer than 25% are highly fruitful. The berries on most MIKs average under 1g in weight, but are much larger than those of *V. darrowi, V. arboreum* or the F1 hybrids. Fruit color ranges from medium blue to black. Scars are often poor, with tears at the pedicel attachment point. None of the MIKs produce berries of cultivar quality, but many are vigorous, flower heavily, and appear more tolerant of unamended low-organic soils than southern highbush seedlings.

Ten of the MIKs that were most fruitful after open-pollination in the field, produced numerous seedlings after pollination in a greenhouse with pollen from southern highbush cultivars. This indicated that the fertile MIKs were mostly tetraploid or near-tetraploid and probably originated by union of unreduced megagametes from the diploid F1 hybrids with normal microgametes from tetraploid southern highbush.

Characteristics of MIK × Southern Highbush Seedlings

Vigor of seedlings from the MIK × southern highbush cultivar crosses varied greatly from cross to cross. In about half of the 16 crosses, seedling vigor averaged low. For several crosses, all seedlings were weak and chlorotic after 2 years in a field nursery. Other crosses produced highly vigorous seedlings. No reason was apparent for the variability among crosses.

Within the vigorous populations, most seedlings flowered heavily and had high percent fruit set after open-pollination in the field. Some seedlings, however, flowered but set little or no fruit. Time of fruit ripening varied greatly among seedlings within crosses. Some seedlings ripened about the same time as the earliest southern highbush cultivars, with ripe fruit by April 15, 1994 in Gainesville. Other seedlings began ripening more than a month later. The 1994 ripening season for blueberries in north Florida was about 2 weeks ahead of normal due to above-normal temperatures from February through April.

Fruit quality on the best backcross seedlings was much improved over that of the MIKs. Although berries were mostly too small for cultivars, and bad scars were a common defect, fruit on a few of the seedlings was at or near cultivar quality. Fruit color was surprisingly good for many of the seedlings, and pedicels on many were unusually long.

The approximate genetic composition of the backcross seedlings is thought to be one-eighth *V. arboreum*, one-eighth *V. darrowi* derived from the F1 hybrid parent, and three-quarters southern highbush, which is mostly northern highbush and *V. darrowi*. It is too early to determine whether any of the backcross seedlings with high fruit quality inherited useful levels of sparkleberry tolerance for low-organic, high-pH soil. Many appear to be highly vigorous in high-density fruiting nurseries.

Characteristics of MIK × MIK
and MIK Self-Pollinated Seedlings

Only MIKs previously selected for high fertility were used in these crosses. As discussed above, many MIKs had little or no fertility. The genetic composition of the progeny from these crosses is believed to be about one-quarter *V. arboreum*, one-quarter *V. darrowi* derived from the F1 *V. darrowi* × *V. arboreum* parent, and half southern highbush. The MIK × MIK seedlings were highly variable in vigor, fruit quality, and time of ripening. Some ripened their first berries about June 1, 1994 in the field in Gainesville, two weeks after the latest southern highbush were 100% ripe and at a time when the midseason rabbiteye varieties 'Tifblue' and 'Brightwell' were 50% ripe. Fruit set was mostly high on seedlings from MIK × MIK crosses when they were allowed to open-pollinate in the field.

Plant vigor and growth habit were highly variable, both within and among MIK × MIK seedlings populations. Vigor averaged about the same as for the MIK × southern highbush backcrosses, but there was more variability within crosses and less among crosses for the MIK × MIK. The fruit was relatively small, averaging below 1 g per berry for most plants, and for many plants, the skin tore at the pedicel attachment point when the fruit was picked. Fruit color was surprisingly good for many seedlings, approaching the powdery blue of the best cultivars. The high variability of the MIK × MIK seedlings would allow selection of superior plants for use in backcrosses to highbush.

Seedlings from self-pollination of the most fertile MIKs averaged lower in vigor than MIK × MIK seedlings, as would be expected based on the inbreeding depression previously observed in blueberries (Lyrene, 1983). A few selfed seedlings, however, were highly vigorous and very late ripening. Except for these few anomalous seedlings, selfing the MIKs did not seem to be a very useful breeding strategy.

GROWER BENEFITS

Highbush blueberry production is limited to a few soil types. These are unavailable in many areas. The moist, organic soils favored by highbush blueberries are usually located in low-lying, frost-prone areas, and are sometimes hard to manage because they have a high and fluctuating water table. The sparkleberry tree can grow on upland soils that have little water-holding capacity and on soils too alkaline for highbush blueberries. It is heat tolerant and resistant to various diseases that afflict highbush blueberries. The sparkleberry flowers late in the spring, escaping freezes that often damage the highbush blueberry crop, and its short, widemouthed flowers are easier for some bee species to pollinate than the longer, narrower flowers of the highbush blueberry. This paper shows that sparkleberry can be used in breeding highbush blueberry varieties. Plants that contain as much as one-eighth sparkleberry parentage produce berries that meet commercial standards for highbush blueberry. Further breeding with sparkleberry may produce commercial highbush cultivars that are more widely adapted and easier to grow.

LITERATURE CITED

Ballington, J.R., B.W. Foushee, and F. Williams-Rutkosky. Potential of chip-budding, stub-grafting, or hot-callusing following saddle-grafting on the production of grafted blueberry plants. Proc. of 6th North American Blueberry Res. Workers Conference. Portland, Oregon. pp. 114-120.

Camp, W.H. 1945. The North American blueberries with notes on the other groups of Vacciniaceae. Brittonia 5:203-275.

Coville, F.V. 1937. Improving the wild blueberry. pp. 559-574. In 1937 Yearbook of Agriculture, United States Government Printing Office, Washington, DC.

Luby, J.L., J.R. Ballington, A.D. Draper, K. Pliszka, and M.E. Austin. 1990. Blueberries and cranberries *(Vaccinium)*. pp. 393-456. In: Genetic Resources of Temperate Fruit and Nut Crops. Vol. 1. Ed. by J.N. Moore and J.R. Ballington, Jr. Int. Society for Hort. Sci. Wageningen.

Lyrene, P.M. 1983. Inbreeding depression in rabbiteye blueberries. HortScience 18:226-227.

Lyrene, P.M. 1991. Fertile derivatives from sparkleberry × blueberry crosses. J. Amer. Soc. Hort. Sci. 116:899-902.

Sharpe, R.H. 1954. Horticultural development of Florida blueberries. Proc. Fla. State Hort. Soc. 66:188-190.

Sharpe, R.H. and G.M. Darrow. 1959. Breeding blueberries for the Florida climate. Proc. Fla. State Hort. Soc. 72:308-311.

Sharpe, R.H. and W.B. Sherman. 1971. Breeding blueberries for low-chilling requirement. HortScience 6:145-147.

Sherman, W.B., R.H. Sharpe, and J. Janick. 1973. The fruiting nursery: Ultra high density for evaluation of blueberry and peach seedlings. HortScience 8:170-172.

Stockton, Jr., L.A. 1976. Propagation and autecology of *Vaccinium arboreum* and its graft compatibility with *V. ashei*. M.S. Thesis, Texas A&M Univ., College Station, TX.

Progress Toward Identifying Markers Linked to Genes Controlling Chilling Requirement and Cold Hardiness in Blueberry

Lisa J. Rowland
Amnon Levi
Rajeev Arora
Elizabeth L. Ogden
Mubarack M. Muthalif
Nicholi Vorsa
Richard G. Novy
Michael E. Wisniewski

SUMMARY. Progress was made in identifying molecular markers linked to genes which control chilling requirement and cold hardiness in a woody perennial, blueberry (*Vaccinium*, section *Cyanococcus*).

Lisa J. Rowland and Amnon Levi, Fruit Laboratory, Agricultural Research Service, U.S.D.A., Beltsville, MD 20705.

Rajeev Arora, Division of Plant and Soil Sciences, West Virginia University, Morgantown, WV 26506.

Elizabeth L. Ogden, Fruit Laboratory, Agricultural Research Service, U.S.D.A., Beltsville, MD 20705.

Mubarack M. Muthalif, Department of Horticulture, University of Maryland, College Park, MD 20742.

Nicholi Vorsa and Richard G. Novy, Blueberry and Cranberry Research Center, Rutgers University, New Jersey Agricultural Experiment Station, Chatsworth, NJ 08019.

Michael E. Wisniewski, Appalachian Fruit Research Station, Agricultural Research Service, U.S.D.A., Kearneysville, WV 25430.

[Haworth co-indexing entry note]: "Progress Toward Identifying Markers Linked to Genes Controlling Chilling Requirement and Cold Hardiness in Blueberry." Rowland, Lisa J. et al. Co-published simultaneously in *Journal of Small Fruit & Viticulture* (Food Products Press, an imprint of The Haworth Press, Inc.) Vol. 3, No. 2/3, 1995, pp. 39-52; and: *Blueberries: A Century of Research* (ed: Robert E. Gough, and Ronald F. Korcak) Food Products Press, an imprint of The Haworth Press, Inc., 1995, pp. 39-52. Single or multiple copies of this article are available from The Haworth Document Delivery Service [1-800-342-9678, 9:00 a.m. - 5:00 p.m. (EST)].

Construction of RAPD (randomly amplified polymorphic DNA)-based
genetic linkage maps using two testcross populations derived from
interspecific hybrids of wild diploid blueberry species, *V. darrowi*
Camp × *V. elliottii* Chapm. and *V. darrowi* × *V. caesariense* Mack-
enz. was continued. Mapping populations were segregating for chilling requirement and cold hardiness. Screening plants of the *V. darrowi* × *V. caesariense*-derived mapping population for cold
acclimation capacity using leaf tissue collected before and after cold
treatment and an electrolyte leakage assay was begun. In addition, a
tetraploid breeding population (predominantly *V. corymbosum* L.)
was screened for chilling requirement and efforts were begun to
identify markers linked to genes determining chilling requirement using
this population and a bulked-segregant-type analysis. *[Article copies
available from The Haworth Document Delivery Service: 1-800-342-9678.]*

KEYWORDS: Dormancy; Cold acclimation; Freezing tolerance;
Genetic map; RAPD markers; *Vaccinium*

INTRODUCTION

To survive the winters, woody perennial plants of the temperate
zone must enter a state of dormancy and develop freezing tolerance
(Powell, 1987). Exposure to low temperatures, which plays a role in
the development of freezing tolerance, is also required for breaking
dormancy and resumption of growth the following spring (Scalabrelli and Couvillon, 1986). This requirement, called the chilling
requirement or CR, is genetically determined (Samish, 1954;
Hauagge and Cummins, 1991). CR is calculated slightly differently
depending on the model used; one example is the number of hours
between 0 and 7°C (32 and 44.6°F) necessary for >50% budbreak
upon exposure to temperatures near 20°C (68°F). The CR is a
critical factor controlling the life cycle of a woody plant, serving to
synchronize a plant's growth with environmental conditions.

A goal of our research is to identify molecular markers linked to
genes which control CR and cold hardiness or CR in a woody perennial, blueberry (*Vaccinium*, section *Cyanococcus*), because blueberries with broader climatic adaptation are of particular interest in
blueberry breeding programs today. One of the emphases in the
USDA blueberry breeding program in Beltsville, Maryland for the

past couple of decades was to develop hybrid highbush cultivars (*V. corymbosum* × *V. darrowi*) suitable for growing in the southern United States because they are generally earlier ripening than the rabbiteye (*V. ashei* Reade) cultivars grown there. This led to the release of several low chilling predominantly highbush cultivars like 'Gulfcoast,' 'Georgiagem' and 'Cooper' (Austin and Draper, 1987). An emphasis of other blueberry breeding programs, in the state of Minnesota, for example, has been to develop more cold hardy cultivars (Fear et al., 1985; Finn and Luby, 1990). This has been achieved by hybridizing lowbush and highbush species to produce "half-high" types (*V. angustifolium* × *V. corymbosum*) that perform well in regions like Minnesota and northern Michigan.

If molecular markers linked to genes controlling CR and CH were found, they could be followed in breeding populations to identify plants, while still seedlings, with the appropriate CR or level of CH. To achieve our goal of identifying these markers, we are currently (1) constructing RAPD (randomly amplified polymorphic DNA)-based genetic linkage maps of two testcross populations segregating for CR and CH derived from interspecific hybrids of wild diploid species, *V. darrowi* × *V. elliottii* and *V. darrowi* × *V. caesariense*, (2) screening individuals of the *V. darrowi* × *V. caesariense* population for ability to acclimate to cold in an attempt to identify markers linked to CH genes, and (3) performing a bulked-segregant-type analysis (Michelmore et al., 1991) using a tetraploid breeding population screened for CR to identify markers linked to genes controlling CR. Here we report our progress in each of these areas.

MATERIALS AND METHODS

Plant Material: One of our mapping populations resulted from a testcross between one interspecific hybrid US388 (*V. darrowi* clone Fla4B [evergreen, low chilling lowbush] × *V. elliottii* clone Knight [deciduous, moderately high chilling highbush]) and another *V. darrowi* clone, US799. The other mapping population resulted from testcrosses between three interspecific hybrids 141-5, 141-6, and 141-10 (*V. darrowi* clone Fla4B × *V. caesariense* clone W85-20 [deciduous, high chilling highbush]) and other *V. darrowi* and *V. caesariense* clones, NJ88-13-15 and W85-23, respectively. Fla4B, US799, and NJ88-13-15 are wild selections collected in Florida. Knight is a wild selection collected in North Carolina. W85-20 and W85-23 are wild selections from New Jersey. The tetraploid breeding population used for CR determinations (see below) resulted from a backcross between a tetraploid interspecific hybrid US340 (*V. darrowi* clone Fla4B [diploid] × *V. corymbosum* cv. Bluecrop [tetraploid]) and 'Bluecrop.' Fla4B, Knight, US388, US799, and US340 were kindly provided by A. Draper.

RAPD Reactions: DNA was extracted from leaf tissue from plants of the three populations (described above) using a CTAB (hexadecyltrimethylammonium bromide) procedure (Doyle and Doyle, 1987) as modified by Rowland and Nguyen (1993) but with the omission of the polyethylene glycol precipitation. Amplification reactions were performed as described previously (Levi et al., 1993; Rowland and Levi, 1994) using 10-base primers with GC contents ranging from 50-100% (Biotechnology Laboratory, University of British Columbia). Amplification products were separated by electrophoresis through 1.4% agarose gels in 0.5 × Tris-borate electrophoresis buffer (Sambrook et al., 1989). DNA fragments were visualized by UV irradiation after staining with 0.5 μg/ml ethidium bromide.

Freezing Hardiness Determinations: Leaf tissue was collected from parent plants of our *V. darrowi* × *V. caesariense*-derived mapping population before and after a three-week low temperature (4°C [39.2°F]), short photoperiod (10 hour day/14 hour night) treatment. Freezing tolerance was determined before and after cold treatment as described by Steffen et al. (1989) and Stone et al. (1993) by gradually cooling (1-2°C/hour) the leaf tissue to temperatures ranging from 0 to −15°C. Samples were removed at predetermined temperatures and ion leakage was measured as previously described (Arora et al., 1992). Samples were then heat-killed in the same solution and total conductivity was once again recorded at room temperature. Percent ion leakage (% of total) was plotted as a function of freezing temperatures. Relative freezing tolerance or RFT was calculated for nonacclimated and acclimated samples from the midpoint of the maximum and minimum (control) ion leakage values (Stone et al., 1993).

CR Determinations: Approximately 10-15 cm (4-6 in) long shoots were cut from plants of the US340 × 'Bluecrop' population after various lengths of chilling, placed in jars of water, and held at 20°C (68°F) for three weeks. During the winter of 1992-1993, two shoots were cut from each plant; and during the winter of 1993-1994, five to six shoots were cut from each plant. Shoots were cut after 690, 820, 940, and 1150 chill units (CUs) in 1992-1993 and after 300, 660, 910, and 1320 CUs in 1993-1994. Here CUs are defined as the number of hours of exposure to temperatures from 0-7.2°C (32-45°F). Shoots were rated for floral bud development after three weeks at 20°C (68°F) as described by Spiers (1978). Accordingly, each bud was examined for signs of budbreak and given a score from 1-7, with 1 representing the stage with no sign of visible swelling and 7 representing the stage after which corollas had completely expanded and dropped. The average bud score was then determined for each plant at each time point. The 10-20 plants with the highest average bud score at 660-690 CUs (lowest CR) and those with the lowest average bud score at 1150-1320 CUs (highest CR) were identified for each year. Seven to ten plants of each group that behaved consistently from year to year were chosen for bulked segregant analysis (Michelmore et al., 1991).

RESULTS AND DISCUSSION

Construction of RAPD-Based Linkage Maps: We recently reported the construction of an initial genetic linkage map for diploid blueberry using a population resulting from a testcross between the F_1 interspecific hybrid US388 (*V. darrowi* × *V. elliottii*) and another *V. darrowi* clone, US799 (Rowland and Levi, 1994). The map currently comprises 72 RAPD markers mapped to 12 linkage groups, in agreement with the basic blueberry chromosome number. Four hundred 10-base primers have now been screened in amplification reactions using DNA isolated from the parent plants of the *V. darrowi* × *V. caesariense*-derived population to identify DNA fragments suitable for mapping in this population. DNA fragments suitable for mapping in the whole population must be unique to one of the original parents (Fla4B or W85-20), present in all three F_1s

(141-5, 141-6, and 141-10) and absent from both testcross parents (NJ88-13-15 and W85-23). To date, 25 such markers have been identified. In the process of screening the parent plants of both mapping populations, about 110 more markers were found that are suitable for mapping in the *V. darrowi* × *V. elliottii*-derived population. Of the 25 markers identified for the *V. darrowi* × *V. caesariense* population, about six appeared to be identical to markers previously mapped in the *V. darrowi* × *V. elliottii*-derived population. Presently, segregation ratios of 16 of the 25 markers have been determined for the whole *V. darrowi* × *V. caesariense*-derived population. All segregated at the expected 1:1 ratio (for a typical gel, see Figure 1).

Cold Acclimation Studies: Previously, we have identified three proteins of 65, 60, and 14 kD, responsive to chilling and immuno-

FIGURE 1. Amplification reactions using a 10-base primer from the University of British Columbia and blueberry DNA from the *V. darrowi* × *V. caesariense*-derived mapping population as template. From left to right (lanes A-S) are shown PCR using original parents Fla4B and W85-20, Fla4B × W85-20 F1s (141-5, 141-6, and 141-10), testcross parents W85-23 and NJ88-13-15, and then PCR using a total of 12 individuals of the mapping population. The arrows to the right of the gel point to two segregating RAPD markers.

A B C D E F G H I J K L M N O P Q R S

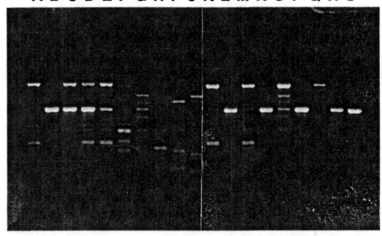

logically-related to dehydrins (Close et al., 1993), that increase in blueberry floral buds during cold acclimation and decrease during deacclimation and resumption of growth (Muthalif and Rowland, 1994) (for a representative protein gel, see Figure 2). Further characterization of these three dehydrin-like proteins indicated that they can be induced in leaves also with two weeks of exposure to 4°C (39.2°F). Experiments are currently underway to (1) identify loci that control cold acclimation ability in our mapping population as already described, (2) isolate cDNA clones encoding the three dehydrin-like proteins using antiserum raised against them, and (3) use the cDNA clones and our mapping population to map the genes encoding these proteins. Then, it will be determined whether these genes co-segregate with loci identified in our mapping population that control the ability to acclimate to cold. If they do not, we will rule out the hypothesis that these proteins play a major causal role in determining cold acclimation ability. If they do co-segregate, we will have strong evidence for a cause-and-effect relationship.

The notion that Fla4B and W85-20, the original parent plants of the *V. darrowi* × *V. caesariense*-derived mapping population, may be significantly different in terms of their cold acclimation ability (based on their collection from Florida and New Jersey, respectively) and, thus, the mapping population should be segregating for cold acclimation ability, has been recently confirmed in our lab. LT_{50} values (temperature at which 50% of the flowers are killed) based on visual observations were determined for floral buds of Fla4B and W85-20 after a two-week treatment at 4°C (39.2°F). LT_{50} for Fla4B was −7°C (19.4°F) whereas it was three times lower or −21°C (−5.8°F) for W85-20 (data not shown). Because at this time all individuals of the mapping population are not mature enough to have set flower buds and because we now know the three dehydrin-like proteins, originally found in blueberry floral buds, are inducible in leaves, this year we have explored the possibility of screening the mapping population for cold acclimation ability using leaf tissue rather than floral bud tissue. Leaf tissue was collected from parent plants before and after a three-week treatment at 4°C (39.2°F) and freezing tolerance determined using an electrolyte leakage assay (Stone et al., 1993). Relative freezing tolerance (RFT) of the parent plants in the nonacclimated and cold acclimated states

FIGURE 2. Profiles of soluble proteins extracted from floral buds of *V. corymbosum* cv. Bluecrop (CR of 1200 CUs) collected at various times during CU accumulation until the resumption of growth. Proteins were extracted and equal amounts of proteins were loaded from each time point and separated on a 12.5% gel by SDS-PAGE, all as previously described (Muthalif and Rowland, 1994). CUs are indicated by the number above each lane. In the far left lane are molecular weight markers. Arrows to the right mark the 65-, 60-, and 14-kD polypeptides that accumulate with chilling and decrease with the resumption of growth. In this case, buds had begun to swell by about 1800 CUs.

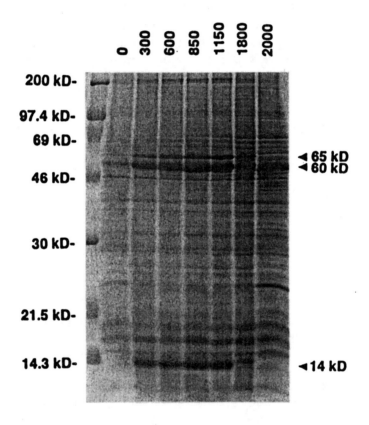

TABLE 1. Relative freezing tolerance (RFT) for nonacclimated and cold acclimated leaves of original parent plants, testcross parent plants, and F₁s of the *V. darrowi* × *V. caesariense*-derived mapping population.

	Relative Freezing Tolerance (°C)*	
Plant	Nonacclimated	Cold Acclimated
Fla4B	− 6.4	− 8.4
W85-20	− 6.0	− 10.5
W85-23	− 9.2	− 9.5
141-5	− 5.5	− 8.4
141-6	− 5.4	− 6.0
141-10	− 7.2	− 9.5

*RFT was determined from the midpoint of the maximum and minimum (nonfrozen control) ion leakage values obtained for each genotype.

(as described under "Materials and Methods") is presented in Table 1. At the time of writing this report, RFT of the *V. darrowi* testcross parent NJ88-13-15 had not been determined because too little leaf tissue was available for this analysis. Although the RFT of the cold acclimated *V. darrowi* parent Fla4B was lower than the RFTs of the cold acclimated *V. caesariense* parents, W85-20 and W85-23, the differences were relatively small (2.1 and 1.1°C, respectively). In addition, the RFTs of the F₁s ranged from below that of the *V. darrowi* parent Fla4B to nearly the same as that of the *V. caesariense* parents. For these reasons, we have concluded that the mapping population should be screened for ability to cold acclimate at a later date using floral bud tissue instead of leaf tissue.

Chilling Requirement Studies: Because, as stated above, all individuals of our mapping populations have not set flower buds, we have begun efforts to identify markers linked to CR genes using a mature tetraploid breeding population. This backcross population is being used in conjunction with a bulked-segregant-type analysis

(Michelmore et al., 1991) to identify the markers. The bulked-segregant analysis involved first screening plants of the population to identify those with the lowest and highest CRs (see Figure 3 for the distribution of CRs of the backcross progenies). Then, DNA was isolated from seven to eight plants of each of the extreme phenotypes. Equal amounts of DNA from each individual plant of the low-chilling type was combined or "bulked" as was DNA from plants of the high-chilling type. These bulks are currently being screened in RAPD reactions using many different primers in an attempt to identify one or more markers present in one of the bulks but absent in the other. If a putative marker is found, reactions will be set up using DNA from each of the seven to eight plants of each extreme phenotype separately to confirm the presence of the marker in all individuals of one extreme and the absence of the marker in all individuals of the other extreme. Segregation of a marker with the high or low CR phenotype would indicate linkage of the marker with a gene(s) controlling this trait. Presently, approximately 20 primers have been used in RAPD reactions along with DNA from the parent plants, US340 and 'Bluecrop,' and the bulked DNAs. To date, one marker has been identified which putatively co-segregates with the low CR phenotype. This marker is present in US340 and the low CR bulk and absent in 'Bluecrop' and the high CR bulk (see Figure 4).

CONCLUSIONS

An objective of our research is to identify molecular markers linked to genes which control CR and CH in blueberry. To achieve this goal, RAPD-based genetic linkage maps are being constructed using two diploid populations segregating for CR and CH. To date, 72 RAPD markers have been mapped in the first *V. darrowi* × *V. elliottii*-derived population (Rowland and Levi, 1994) and another 25 RAPD markers have been identified suitable for mapping in the second *V. darrowi* × *V. caesariense*-derived population. Segregation of 16 of the 25 markers has been examined in the entire population. Screening this same *V. darrowi* × *V. caesariense*-derived population for cold acclimation ability using leaf tissue collected from plants before and after a cold treatment was explored. However,

FIGURE 3. Distribution of average floral bud scores among the US340 ×
'Bluecrop' population. Number of plants were plotted on the y axis versus
average bud score plotted on the x axis. Average bud scores were calculated as described in the text. Shown are data from 1993-1994 at 660 (A) and
1320 CUs (B). The plants with the highest average bud scores at 660 CUs
were considered to have the lowest CRs of the population whereas the
plants with the lowest average bud scores at 1320 CUs were considered to
have the highest CRs of the population.

A. 660 CUs

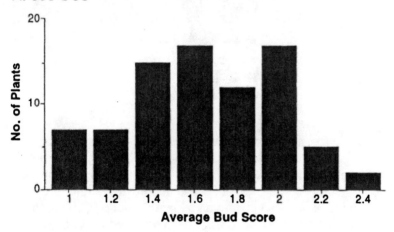

B. 1320 CUs

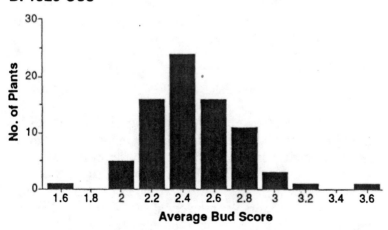

FIGURE 4. Amplification reactions using one 10-base primer and DNA from US340, 'Bluecrop,' low CR bulk, and high CR bulk (lanes A-D) as templates. The arrow points to the segregating marker present in reactions using US340 and the low CR bulk as template and absent in reactions using 'Bluecrop' and the high CR bulk as template.

A B C D

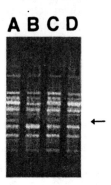

differences in freezing tolerance between parent plants appeared too small using leaf tissue, as compared to floral bud tissue, to justify this approach. Because all individual plants of this population have not yet set flower buds, plants of a mature tetraploid breeding population are being used now to also identify markers linked to CR genes. Plants of this population have been screened for CR over two consecutive years. Plants with the lowest and highest CRs have been identified and are being used in a bulked-segregant analysis (Michelmore et al., 1991) to identify markers linked to genes controlling CR. One marker has been found which putatively co-segregates with the low CR phenotype.

GROWER BENEFITS

The identification of markers linked to genes which control CR and CH in blueberry would be useful for breeding blueberry plants with greater climatic adaptation. Plants could be evaluated at the seedling stage for the likelihood of having a low or high CR or level of CH by the presence or absence of specific DNA markers. Thus, seedlings having less desirable traits could be discarded rather than being grown to maturity saving labor, time, and space.

LITERATURE CITED

Arora, R., M.E. Wisniewski, and R. Scorza. 1992. Cold acclimation in genetically related (sibling) deciduous and evergreen peach (*Prunus persica* [L.] Batsch). Plant Physiology 99:1562-1568.

Austin, M.E and A.D. Draper. 1987. 'Georgiagem' blueberry. HortSci. 22:682-683.

Close, T.J., R.D. Fenton, A. Yang, R. Asghar, D.A. DeMason, D.E. Crone, NC. Meyer, and F. Moonan. 1993. Dehydrin: the protein. pp. 104-118. In: T.J. Close and E.A. Bray (eds.), *Plant responses to cellular dehydration during environmental stress.* American Society of Plant Physiologists, Rockville, MD.

Doyle, J.J. and J.L. Doyle. 1987. A rapid DNA isolation procedure for small quantities of fresh leaf tissue. Phytochem. Bull. 19:11-15.

Fear, C.D., F.I. Lauer, J.J. Luby, and R.L. Stucker. 1985. Genetic components of variance for winter injury, fall growth cessation, and off-season flowering in blueberry progenies. J. Amer. Soc. Hort. Sci. 110:262-266.

Finn, C.E. and J.J. Luby. 1990. Halfhigh blueberry cultivars. Fruit Varieties Journal 44:63-68.

Hauagge, R. and J.N. Cummins. 1991. Genetics of length of dormancy period in *Malus* vegetative buds. J. Amer. Soc. Hort. Sci. 116:121-126.

Levi, A., L.J. Rowland, and J.S. Hartung. 1993. Production of reliable randomly amplified polymorphic DNA (RAPD) markers from DNA of woody plants. HortSci. 28:1188-1190.

Michelmore, R.W., I. Paran, and R.V. Kesseli. 1991. Identification of markers linked to disease-resistance genes by bulked segregant analysis: a rapid method to detect markers in specific genomic regions. Proc. Natl. Acad. Sci. USA 88:9828-9832.

Muthalif, M.M. and L.J. Rowland. 1994. Identification of dehydrin-like proteins responsive to chilling in floral buds of blueberry (*Vaccinium*, section *Cyanococcus*). Plant Physiol. 104:1439-1447.

Powell, L.E. 1987. Hormonal aspects of bud and seed dormancy in temperate-zone woody plants. HortSci. 22:845-850.

Rowland, L.J. and A. Levi. 1994. RAPD-based genetic linkage map of blueberry derived from a cross between diploid species (*Vaccinium darrowi* and *V. elliotii*). Theor. Appl. Genet. 87:863-868.

Rowland, L.J. and B. Nguyen. 1993. Use of PEG for purification of DNA from leaf tissue of woody plants. BioTechniques 14:735-736.

Sambrook, J., F.F. Fritsch, and T. Maniatis. 1989. *Molecular cloning: a laboratory manual.* Cold Spring Harbor Press, Cold Spring Harbor, NY.

Samish, R.M. 1954. Dormancy in woody plants. Annu. Rev. Plant Physiol. 5:183-204.

Scalabrelli, G. and G.A. Couvillon. 1986. The effect of temperature and bud type on rest completion and the GDH°C requirement for budbreak in 'Redhaven' peach. J. Amer. Soc. Hort. Sci. 111:537-540.

Spiers, J.M. 1978. Effect of stage of bud development on cold injury in rabbiteye blueberry. J. Amer. Soc. Hort. Sci. 103:452-455.

Steffan, K.L., R. Arora, and J.P. Palta. 1989. Relative sensitivity of photosynthesis and respiration to freeze-thaw stress in herbaceous species: importance of realistic freeze-thaw protocols. Plant Physiol. 89:1372-1379.

Stone, J.M., J.P. Palta, J.B. Bamberg, L.S. Weiss, and J.F. Harbage. 1993. Inheritance of freezing resistance in tuber-bearing *Solanum* species: evidence for independent genetic control of nonacclimated freezing tolerance and cold acclimation capacity. Proc. Natl. Acad. Sci. USA 90:7869-7873.

The Search for Chilling-Responsive Proteins in Blueberry Continues

Mubarack M. Muthalif
Lisa J. Rowland

SUMMARY. Previously, we identified three proteins in blueberry floral buds of 65, 60, and 14 kD that increase dramatically in abundance during chill-unit accumulation and cold acclimation and decrease during deacclimation and resumption of growth. These proteins are related immunologically to a group of proteins called dehydrins. Here we report results from immunoblot experiments using antiserum raised against molecular chaperones, CAP 79 from spinach and a cyclophilin from *Arabidopsis*, in addition to dehydrin antiserum. Using these antisera, two new chilling-responsive polypeptides have been identified in blueberry floral buds. One is a 22 kD polypeptide immunologically related to dehydrins and the other is a 17.5 kD polypeptide immunologically related to cyclophilins. The 17.5 kD polypeptide was purified and partially sequenced. Sequence information confirmed the 17.5 kD polypeptide is, indeed, cyclophilin and is very similar in sequence to cyclophilins identified in other plants. *[Article copies available from The Haworth Document Delivery Service: 1-800-342-9678.]*

KEYWORDS: Dormancy; Chilling requirement; Cold hardiness; Cold acclimation; Dehydrins; Cyclophilins; HSPs; *Vaccinium*

Mubarack M. Muthalif, Department of Horticulture, University of Maryland, College Park, MD 20742.

Lisa J. Rowland, Fruit Laboratory, Agricultural Research Service, U.S.D.A., Beltsville, MD 20705.

[Haworth co-indexing entry note]: "The Search for Chilling-Responsive Proteins in Blueberry Continues." Muthalif, Mubarack M., and Lisa J. Rowland. Co-published simultaneously in *Journal of Small Fruit & Viticulture* (Food Products Press, an imprint of The Haworth Press, Inc.) Vol. 3, No. 2/3, 1995, pp. 53-60; and: *Blueberries: A Century of Research* (ed: Robert E. Gough, and Ronald F. Korcak) Food Products Press, an imprint of The Haworth Press, Inc., 1995, pp. 53-60. Single or multiple copies of this article are available from The Haworth Document Delivery Service [1-800-342-9678, 9:00 a.m. - 5:00 p.m. (EST)].

INTRODUCTION

To survive the winters, woody perennial plants have evolved a mechanism by which they become dormant and cold hardy in the autumn. Although development of dormancy and development of cold hardiness or CH (cold acclimation or CA) are integral parts of their life cycle, little research has been done to investigate changes in levels of proteins in response to chill-unit (CU) accumulation or low temperature stress in woody perennials. For this reason, we have been using blueberry (*Vaccinium* section *Cyanococcus*) plants as a system to investigate induction of specific proteins in floral buds of woody perennials during CU accumulation and CA. We have identified three proteins in blueberry floral buds of 65, 60, and 14 kD that increase dramatically in abundance during CU accumulation and CA and decrease during deacclimation and resumption of growth (Muthalif and Rowland, 1994a,b). These proteins are related immunologically to a group of proteins previously found in annual plants called dehydrins that are induced during drought and low temperature stress (Close et al., 1993).

Members of another group of proteins called molecular chaperones (Thomashow, 1993) have been identified in some annual plants as being responsive to low temperature stress as well. The primary function of molecular chaperones is to assist in the transport, folding, and assembly of other proteins (Thomashow, 1993). Examples of cold-induced molecular chaperones that have been identified in other plants are the CAP 79 polypeptide from spinach, which is a member of the HSP 70 family of proteins (Nevin et al., 1992), and cyclophilin from maze (Marivet et al., 1992). Consequently, we have used antiserum raised against CAP 79 and a cyclophilin from *Arapidopsis* to determine whether immunologically-related proteins are induced in floral buds of blueberry during CU accumulation. Results from these experiments are presented below.

MATERIALS AND METHODS

Plant Material: Blueberry genotypes with high ('Bluecrop'–1200 CUs), medium ('Berkeley'–900 CUs), and low ('Tifblue'–600

CUs) chilling requirements or CRs were used. 'Bluecrop' and 'Berkeley' are highbush *V. corymbosum* L. cultivars; 'Tifblue' is a rabbiteye *V. ashei* Reade cultivar. Beginning in October, floral buds were collected about every 300 CUs from field-grown plants of 'Bluecrop' and 'Berkeley' at the Blueberry/Cranberry Research Station (Chatsworth, NJ) and from field-grown plants of 'Tifblue' at the Agricultural Research Center (Beltsvllle, MD) until the resumption of growth in the spring. After collection, buds were frozen in liquid N_2 and stored at $-70°C$ until analyzed. CUs, calculated using either a biophenometer (Omnidata, Logan, UT) or weather data, were defined as the number of hours that plants were exposed to temperatures from 0 to 7°C.

Protein Extractions and Gel Electrophoresis: Soluble proteins were extracted using a phenol-based procedure that lacked SDS in the extraction buffer (Shao-bing et al., 1989). Proteins were quantified using a protein assay kit (BioRad, Richmond, CA) based on the Bradford (1976) method. Proteins from buds exposed to different chilling regimes were separated by SDS-PAGE (Laemmli, 1970) using 12.5% gels.

Immunoblots: Polypeptides were electroblotted onto nitrocellulose membranes and immunostained with the appropriate antiserum. Antisera used were dehydrin (T. J. Close, University of California, Riverside), CAP 79 (C. L. Guy, University of Florida), and cyclophilin (C. S. Gasser, University of California, Davis) antisera. Blots were immunostained using a streptavidin-alkaline phosphatase immunoblotting kit for detecting rabbit or mouse antibodies as appropriate (Gibco BRL, Gaithersburg, MD) and the dilution of antiserum recommended by the laboratory from which the antiserum came. Dilutions were 1:400, 1:10,000, and 1:5000 for the dehydrin, CAP 79, and cyclophilin antisera, respectively.

Purification of Blueberry Cyclophilin, Cleavage, and Sequencing: Soluble proteins were extracted as described above from floral buds of 'Berkeley' plants that had received 950 CUs. Nine milligrams of protein were fractionated by preparative, free-solution isoelectric focusing using the Rotofor (BioRad) as described previously (Muthalif and Rowland, 1994a). The Rotofor fraction (fraction #19) containing the 17.5 kD polypeptide immunologically-related to a cyclophilin from *Arabidopsis* was concentrated, separated by SDS-PAGE, and electroblotted onto a nitrocellulose membrane. After transfer, the polypeptides were reversibly stained with Ponceau S using the method described by Aebersold et al. (1987). The protein band corresponding to the 17.5 kD polypeptide was cut out and shipped to the Harvard Microchemistry Department for sequencing. There the polypeptide was first digested with trypsin and the resulting peptides were separated on a narrow-bore reverse-phase HPLC system (Aebersold et al., 1987). Peptide fractions were collected based on A_{210} and two peaks were selected for sequencing. Peptides were sequenced in a gas-phase sequenator.

RESULTS AND DISCUSSION

Previously, using 1-D SDS-PAGE gels stained with Coomassie blue, we identified three polypeptides in blueberry floral buds of 65, 60, and 14 kD that increase in abundance during CU accumulation and CA and decrease during deacclimation and resumption of growth (Muthalif and Rowland, 1994a,b). These proteins were shown to be immunologically related to a group of proteins known as dehydrins. Furthermore, immunoblots using dehydrin antiserum revealed the presence of a fourth polypeptide of 22 kD that reacted specifically to the dehydrin antiserum and increased in abundance with CA and decreased with deacclimation (see Figure 1). This polypeptide was not observed on Coomassie blue-stained gels alone.

To determine if another group of cold-induced proteins (different from dehydrins) known as molecular chaperones could be identified in blueberry floral buds, protein blots were immunostained with antiserum to the CAP 79 polypeptide from spinach (a member of the HSP 70 family of proteins) and with antiserum to cyclophilin from *Arabidopsis* (a member of the immunophilin family of proteins) (see Figures 2 and 3). CAP 79 antiserum reacted to a 70 kD polypeptide in blueberry floral buds; however, abundance of this polypeptide did not appear to change with CU accumulation. Cyclophilin antiserum reacted to a 17.5 kD polypeptide which, on the other hand, did appear to increase slightly in abundance with CU accumulation in floral buds of three different cultivars tested. A

FIGURE 1. Induction of dehydrins in blueberry floral buds with CU accumulation. A blot of soluble proteins extracted from floral buds of 'Tifblue,' 'Berkeley,' and 'Bluecrop' (TF, BK, and BC) was probed with dehydrin antiserum. Buds were collected after different hours of chilling (0, 650, or 850 CUs) as shown above each lane. "L" refers to the molecular weight ladder. Equal amounts of protein were loaded in each lane. The 65, 60, and 22 kD polypeptides are indicated to the left of the blot.

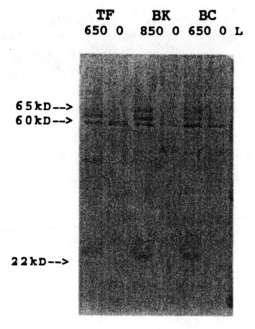

densitometric scan of this blot revealed that the 17.5 kD polypeptide increased by 1.9, 2.8, and 1.3 times in 'Bluecrop,' 'Berkeley,' and 'Tifblue', respectively, with CU accumulation.

The 17.5 kD polypeptide was purified from floral buds of 'Berkeley' and digested with trypsin. Two tryptic fragments were sequenced and the amino acid sequences were used to search the gene bank library for similar sequences (see Figure 4). The first fragment had 88.9, 72.2, and 61.1% similarity to sequences in tomato, *Brassica*, and maize cyclophilins, respectively. The second fragment had 100, 95.5, 90.9, and 81.8% sequence similarity to

FIGURE 2. Response of HSP 70-like proteins to CU accumulation in blueberry floral buds. A blot of soluble proteins extracted from floral buds of 'Bluecrop' was probed with CAP 79 antiserum. Buds were collected after 0, 300, 600, 900, 1200, 1800, and 2000 CUs (lanes 1-7, respectively). "L" refers to the molecular weight ladder. Equal amounts of protein were loaded in each lane. The position of molecular weight markers and the 70 kD immunoreactive product are indicated.

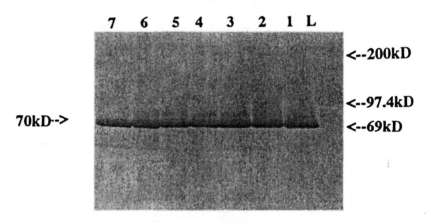

tomato, *Brassica*, maize, and *Arabidopsis* cyclophilins, respectively.

GROWER BENEFITS

The identification of proteins which may be related to development of dormancy or CH in woody perennials could be extremely useful for monitoring levels of dormancy or CH at different stages in a plant's life cycle or in different species or cultivars. Isolation of genes encoding these proteins could open the door to the manipulation of traits such as CR and CH by overexpressing or underexpressing the appropriate genes. Gene probes or antisera to proteins could be developed and used for evaluating breeding populations to identify individual plants with appropriate CRs or level of CH.

FIGURE 3. Induction of cyclophilin in blueberry floral buds with CU accumulation. A blot of soluble proteins extracted from floral buds of 'Tifblue,' 'Berkeley,' and 'Bluecrop' (TF, BK, and BC) was probed with cyclophilin antiserum. Buds were collected after different CUs (0, 650, or 850 CUs) as shown above each lane. "L" refers to the molecular weight markers. Equal amounts of protein were loaded in each lane. The position of molecular weight markers and the 17.5 kD immunoreactive product are indicated.

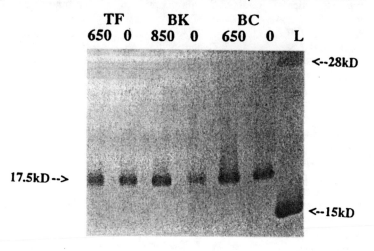

FIGURE 4. Sequence similarity between blueberry cyclophilin and other plant cyclophilins.

SOURCE	AMINO ACID SEQUENCE				SIMILARITY (%)
PEPTIDE I:					
BLUEBERRY BUDS	VTMEL	FADTT	PITAE	NFR	100.0
TOMATO SHOOT TIPS	VVMEL	FADTT	PKTAE	NFR	88.9
BRASSICA BUDS	IVMEL	YADTV	PETAE	NFR	72.2
MAIZE EMBRYO	IVMEL	YANEV	PKTAE	NFR	61.1
PEPTIDE II:					
BLUEBERRY BUDS	VIPGF	MCQGG	DFTAG	NGTGG ES	100.0
TOMATO SHOOT TIPS	VIPGF	MCQGG	DFTAG	NGTGG ES	100.0
BRASSICA BUDS	VIPKF	MCQGG	DFTAG	NGTGG ES	95.5
MAIZE EMBRYO	VIPEF	MCQGG	DFTRG	NGTGG ES	90.9
ARABIDOPSIS	VIPKF	MLQGG	DFTLG	NGRGG ES	81.8

LITERATURE CITED

Aebersold, R.H., J. Leavitt, R.A. Saavedra, L.E. Hood, and S.B.H. Kent. 1987. Internal amino acid sequence analysis of proteins separated by one or two dimensional gel electrophoresis after in situ protease digestion on nitrocellulose. Proc. Natl. Acad. Sci. USA 84:6970-6974.

Bradford, M.M. 1976. A rapid and sensitive method for the quantitation of microgram quantities of proteins utilizing the principle of dye binding. Anal Biochem. 72:248-254.

Close, T.J., R.D. Fenton, A. Yang, R. Asghar, D.A. DeMason, D.E. Crone, N.C. Meyer, and F. Moonan. 1993. Dehydrin: the protein. pp. 104-118. In: T.J. Close and E.A. Bray (eds.), Plant responses to cellular dehydration during environmental stress. American Society of Plant Physiologists, Rockville, MD.

Laemmli, U.K. 1970. Cleavage of structural proteins during the assembly of the head of bacteriophage T_4. Nature 227:680-685.

Marivet, J., P. Frendo, and G. Burkard. 1992. Effects of abiotic stresses on cyclophilin gene expression in maize and bean and sequence analysis of bean cyclophilin cDNA. Plant Science 84:171-178.

Muthalif, M.M. and L.J. Rowland. 1994a. Identification of dehydrin-like proteins responsive to chilling in floral buds of blueberry (*Vaccinium*, section *Cyanococcus*). Plant Physiol. 104:1439-1447.

Muthalif, M.M. and L.J. Rowland. 1994b. Identification of chilling-responsive proteins from floral buds of blueberry. Plant Science, in press.

Neven, L.G., D.W. Haskell, C.L. Guy, N. Denslow, P.A. Klein, L.G. Green, and A. Silverman. 1992. Association of 70-kilodalton heat-shock cognate proteins with acclimation to cold Plant Physiol. 99:1362-1369.

Shao-bing, H., S.K. Dube, N.M. Barnett, and S. Kung. 1989. Polymorphism and Mendelian inheritance of photosystem II 23-kilodalton polypeptide. Planta 179:397-402.

Thomashow, M.F. 1993. Genes induced during cold acclimation in higher plants. pp. 183-210. In: Advances in low-temperature biology. Volume II. JAI Press Ltd.

RELATED POSTERS

Changes in Abscisic Acid and Indoleacetic Acid Levels in 'Climax' Blueberry During Dormancy

J. H. Braswell

SUMMARY. Floral and vegetative plant parts from 'Climax' rabbiteye blueberry *(Vaccinium ashei* Reade) were collected in the fall and winters of 1986-87 and 1989-90 before and after various accumulations of chilling, recorded either as chill units (CU) or chilling hours (CH). Concentrations of abscisic acid (ABA) and indoleacetic acid (IAA) were determined by enzyme-linked immunoassays. Prior to chilling, ABA accumulations were highest in vegetative and flower buds, lowest in the leaves, and intermediate in terminal stems. Vegetative bud ABA content remained high until mid-February when 690

J. M. Spiers is Research Horticulturist, USDA-ARS, Small Fruit Research Station, Poplarville, MS 39470.

J. H. Braswell is Extension Horticulturist, Mississippi Cooperative Extension Service, Poplarville, MS 39470.

[Haworth co-indexing entry note]: "Changes in Abscisic Acid and Indoleacetic Acid Levels in 'Climax' Blueberry During Dormancy." Spiers, J. M., and J. H. Braswell. Co-published simultaneously in *Journal of Small Fruit & Viticulture* (Food Products Press, an imprint of The Haworth Press, Inc.) Vol. 3, No. 2/3, 1995, pp. 61-72; and: *Blueberries: A Century of Research* (ed: Robert E. Gough, and Ronald F. Korcak) Food Products Press, an imprint of The Haworth Press, Inc., 1995, pp. 61-72. Single or multiple copies of this article are available from The Haworth Document Delivery Service [1-800-342-9678, 9:00 a.m. - 5:00 p.m. (EST)].

CH had accumulated. Flower bud ABA content decreased rapidly with minimal CH accumulations (90-380 CH) and then slowly decreased with additional chilling. Individual flower bud dry weight increased 0.03 mg per day from 29 October (CU = 0) to 09 December (CU = 90). Between 09 December and 05 January (CU 450), dormancy was most intense with flower buds increasing only 0.01 mg per day in weight. Flower bud weight increased 0.11 mg per day between 05 January and 20 January (CU = 600). Content of IAA in flower buds decreased between 0 and 90 CU, remained relatively constant between 90 CU and 450 CU and then increased rapidly between 450 and 600 CU collection dates. There was a large increase ($\approx$ 2X) in the ratio of IAA:ABA in the flower buds at the 600 CU date and this corresponded to about a 10-fold increase in daily per bud dry weight gain. These data suggest that IAA levels or the ratio of IAA to ABA may be associated with the regulation of flower bud break in 'Climax' rabbiteye blueberries. *[Article copies available from The Haworth Document Delivery Service: 1-800-342-9678.]*

KEYWORDS: Growth regulators; Rabbiteye blueberry

INTRODUCTION

Blueberries *(Vaccinium* spp.), like many deciduous plants, require a chilling period before vegetative and floral buds break dormancy. Darrow (1942) reported that the rabbiteye blueberry 'Pecan' grew and flowered well after 360 hr of chilling (<7°C). 'Woodard' and 'Tifblue' rabbiteye cultivars required 400 and 600 hrs respectively of constant chilling (6-7°C) for normal flowering (Spiers and Draper, 1974). Austin et al. (1982) reported that rabbiteye blueberry 'Climax' broke floral dormancy after 450 to 550 hrs of chilling while 'Tifblue' and 'Delite' required between 550 and 650 hrs. Additional chilling of rabbiteye blueberries increases the amount and rate of bud break (Spiers and Draper, 1974; Spiers, 1976; Gilreath and Buchanan, 1981; Austin et al., 1982).

The physiological mechanism of the chilling process and subsequent budbreak is not known. One hypothesis in woody plants is that dormancy in buds or seeds is regulated by a balance between growth promoters and growth inhibitors. The concentrations of these plant regulators may be governed by genetic makeup or changes in the environment (Ramsey and Martin, 1970a). Short day-length, which induces dormancy in buds and cessation of growth, resulted in an increase in growth inhibitors and a decrease in growth promoters in the leaves of lamb's-quarter (Nitsch, 1957). In other crops, bud break is associated with variations in growth promoters and growth inhibitors (Nooden and Weber, 1978; Wareing and Saunders, 1971).

Levels of abscisic acid (ABA), a growth inhibitor, declined as dormancy was broken and bud break progressed in a number of fruit species (Corgan and Martin, 1971; Mielke and Dennis, 1975; Wright, 1975; Seeley and Powell, 1981). Reports on tree species indicate that the endogenous levels of growth promoters such as auxins, cytokinins, and gibberellins fluctuate during dormancy and bud break (Bennett and Skoog, 1938; Engelbrecht, 1971; Bachelard and Wrightman,1974; Ramsey and Martin, 1970b).

Alvim et al. (1978) found that long daylength prevented the onset of dormancy in willows (*Salix viminalis* L.) but ABA levels were not significantly changed by photoperiod. He concluded that ABA may not play a role in bud dormancy. Trewavas (1982) stated that tissue sensitivity to growth substances rather than growth substance levels is the governing factor in plant development.

In pecan [*Carya illinoensis* (Wangenh) C. Koch], Wood (1983) found that free and bound ABA levels in apical and basal primary buds declined prior to budbreak until the time of bud break, while IAA levels initially decreased rapidly, then remained low, and finally increased again at budbreak.

An understanding of the activity of endogenous growth regulators during these processes may help in developing cultural methods that regulate fruit production by influencing time of floral bud initiation and break, as well as length of dormant and growing periods. The objective of this study was to determine the levels of a growth promoter, IAA, and a growth inhibitor, ABA, in the floral buds, vegetative buds, stems, and leaves of 'Climax' rabbiteye blueberry during the dormancy period.

MATERIALS AND METHODS

Beginning 29 October 1986, floral buds were collected from mature (8 years old) 'Climax' rabbiteye blueberry plants after accumulations of 0, 90, 450, and 600 chill units (CU). Chill units were measured (Biophenometer Model TA 45) after the maximum negative CU accumulation in the fall according to a modification of the Utah model (Richardson et al., 1974) (Table 1). On each of 4 dates, about 20 buds from each of 20 to 25 plants were collected in bulk and immediately placed in a freezer ($-30°C$). Bud dry ($60°C$) and fresh weight were determined for each sampling date.

In the fall and winter of 1989-90, floral buds (1-5 diallel from terminals), leaves, (3 per terminal), vegetative buds (10 per stem from 1989 stems), and stems (0-3 cm from terminal) were collected from

TABLE 1. Hourly chill unit values recorded per temperature range (calculated at 7.5 minute intervals).

Temperature (°C)	Chill Unit
−50.0	0.00
0.5	0.00
1.5	0.48
2.0	0.63
3.0	0.82
4.5	0.95
6.0	1.00
7.5	0.95
8.5	0.84
9.5	0.70
11.0	0.50
14.5	0.00
21.0	−1.00

the above mentioned 'Climax' plants on 19 October before any chilling accumulated. Chilling hours (hours < 7°C) were recorded instead of chill unit in this study. Additional flower bud, vegetative bud and stem samples were collected on 22 December (380 CU), 12 January (560 CU), 31 January (660 CU), and 14 February (690 CU). Samples were selected from 3 plots (replication), each containing 8 randomly selected plants and were collected from 4 stems per plant on each date.

A modification of Wood's (1983) procedure was used for extraction of plant growth regulators. Frozen floral buds (0.5 g) were ground (Polytron homogenizer) and extracted for 12 hours at $-20°C$ in 20 ml of 80% methanol containing 0.1 mg 2,6-di-tert-butyl-p-cresol (BHT) and 1 mg insoluble polyvinylpyrrolidone (PVP). Five ml of the supernatant was purified through a reverse-phase column (Sep-Pak C-18) for subsequent enzyme immunoassays (EIA). For EIA of free ABA, 1.0 ml of the purified sample extract was diluted 10-fold with 25 mM Tris-buffered saline (TBS) buffer (3.03 g Trizma base, 5.84 g NaCl, and 0.20 g $MgCl_2 \cdot 6H_2O$ per liter H_2O; add 1.0 g NaN_3 and adjust to pH 7.5 with HCl).

For EIA of IAA, 100 µg of purified sample extract was dried with N_2 at 35°C. The dried sample was then reacted with 1 ml of etheral diazomethane. The etheral diazomethane was prepared by first saturating N_2 with diethylether, then passing saturated N_2 through a diazomethane generating solution containing N-methyl-N-nitroso-p-toluenesulfonamide and KOH, and finally passing the gas through a solution of methanol:diethylether (1:9 v/v). The sample was then dried under N_2 at 35°C, dissolved in methanol, diluted with 25 mM Tris buffer.

Sample extracts and standards were dissolved in the Tris buffered solution and subjected to EIA (Phytodetek Kits–Idetek, Inc. San Bruno, CA) for quantitative determination of free ABA and IAA (Weiler, 1980, 1982, 1984; Weiler et al., 1981). Color absorbance (Dynatech Minireader II) was read at 405 nm. The pmoles per sample was determined from a standard curve for each EIA and converted to mg per g (dry weight) of floral buds. Data presented are the means of 2 to 5 extractions per sample and 2 readings per extraction.

RESULTS AND DISCUSSIONS

Concentrations of free ABA declined rapidly from 12.9 to 7.8 ABA per gram bud dry weight as chilling accumulated from 0 to 90 CU (Figure 1). As CU increased from 90 to 600, levels of ABA continued to decline but at decreased rate, from 7.8 to 5.4 µg per gram bud dry weight. Concentration of IAA in floral buds decreased rapidly between 0 and 90 CU, following a pattern similar to ABA (Figure 1). IAA levels remained relatively constant as CU increased from 90 to 450 and then increased slightly as CU accumulated up to 600. IAA concentrations ranged from a low of 104 µg per gram bud dry weight to a high of 258. Free ABA and IAA levels were similar to those reported for buds of several other species (Mielke and Dennis, 1975; Wright, 1975; Wood, 1983).

Floral bud weight increased by 3.1 mg per bud from the 0 to 600 CU accumulations (Figure 2). From 0 to 90 CU, bud dry weight

FIGURE 1. Concentration of free ABA and IAA in floral buds of 'Climax' rabbiteye blueberry as associated with chill unit accumulation.

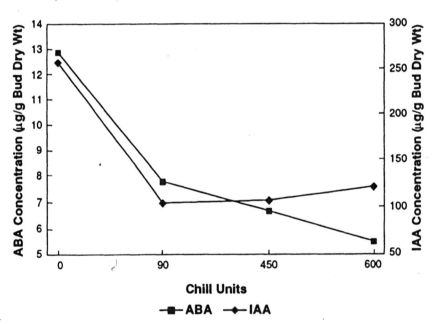

FIGURE 2. Dry weight of 'Climax' rabbiteye blueberry floral buds as associated with chill unit accumulation.

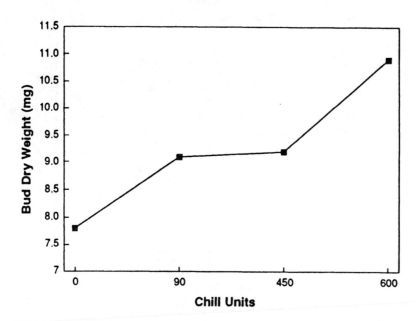

increased from 7.8 to 9.1 mg per bud (0.03 mg per bud per day). Between 90 and 450 CU (05 January), buds averaged only 0.01 mg per day in dry weight gain. Rate of bud weight then increased more rapidly (0.11 mg bud^{-1} day^{-1}) from 450 to 600 CU (20 January), probably indicating a break in dormancy.

On a per bud basis, IAA content decreased from 1380 μg with no chilling to 800 μg with 90 CU (Figure 3) and remained relatively unchanged from 90 to 450 CU. Between 450 and 600 CU, IAA content per bud rapidly increased from 850 to 1260 μg, coinciding with the period of rapid bud weight gain or dormancy cessation. Initially, ABA levels per bud changed similarly to IAA, decreasing from 100 μg ABA with no chilling to 70 μg after 90 CU were recorded. ABA content per bud continued to decline, reaching 60 μg ABA after 600 CU.

As measured by bud dry weight gain per day, the period of most intense bud dormancy was between 90 to 450 CU. During this period IAA per bud remained relatively stable, showing an increase

FIGURE 3. Per bud ABA and IAA content of 'Climax' rabbiteye blueberry floral buds as associated with chill unit accumulation.

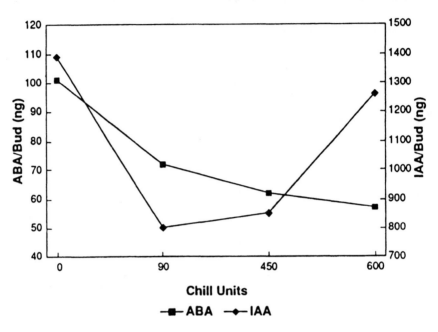

of only 43 µg IAA. Between 450 and 600 CU, the phase of intense bud dormancy appeared to break, as dry bud weight increased at the rate of 0.11 mg per bud per day. These data support results of a previous study by Austin et al. (1982) that indicated that 'Climax' broke dormancy after 450-600 hrs of chilling. The increase in bud weight corresponded to an increase in IAA content per bud of 411 µg. The greatest increase in IAA per bud corresponded with the increase in bud weight which followed the period of most intense bud dormancy.

The IAA:ABA dry weight per bud ratio remained relatively constant for the first 3 chill unit dates and then almost doubled at the last (600 CU) date (Table 2). This large increase in the IAA:ABA ratio corresponded to the 10-fold increase in daily per bud dry weight gain which occurred between the 450 CU and 600 CU collection dates.

In the 1989-90 study, ABA content prior, to chilling (0 CH) was

TABLE 2. Relationship of chill unit accumulations, ratio of IAA:ABA in flower buds, and flower bud dry weights in 'Climax' blueberry.

Chill Units	IAA:ABA ratio (dry wt per bud)	Bud dry wt increase[z] (mg per day)
0	14.3	–
90	11.1	0.03
450	12.5	0.01
600	20.0	0.11

[z]Average daily per bud dry weight increase between individual chill unit collection date listed and previous chill unit collection date (i.e., between 90 CU and 0 CU, 450 CU and 90 CU, and 600 CU and 450 CU).

equally high in vegetative and floral buds (Figure 4). Buds contained about twice as much ABA as the terminal stems; leaves had only a small amount of ABA present. After 380 CH, ABA concentrations in the floral buds and stems had decreased greatly ($\approx 3X$) and then continued to remain at about the same concentrations through the 690 CH collection date. However, vegetative buds responded differently to chilling increases, showing an increase in ABA content after 380 CH and ABA content remained relatively high through the 660 CH collection date (7 February). Only after the 690 CH collection (14 February) did the ABA content decrease rapidly, reaching a level similar to that found in the floral buds. These trends indicate that ABA may play a role in regulating vegetative bud break, which in rabbiteye normally occurs after flowering.

CONCLUSION

Free ABA and IAA levels in floral buds were highest prior to and declined with initial chilling. Levels of ABA were lowest and IAA highest when buds underwent the greatest amount of growth.

FIGURE 4. Concentration of free ABA in floral buds, vegetative buds and stems of 'Climax' rabbiteye blueberry as associated with chilling accumulation.

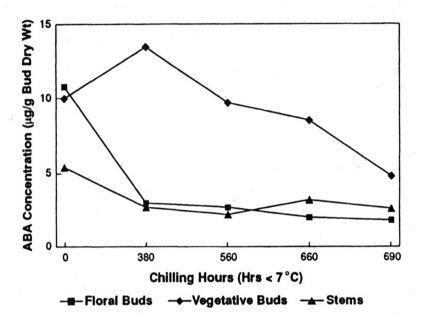

This suggests that free ABA per se may not govern rabbiteye blueberry floral bud dormancy. The rapid decline in IAA levels after accumulation of the first 90 CU, the subsequent low IAA levels, and the eventual sharp rise corresponding to rapid bud weight gain suggest that IAA may be associated with growth activation or a break in dormancy (Bachelard and Wrightman, 1974). Data indicate the operative mechanism for floral budbreak in 'Climax' rabbiteye blueberry may be regulated by a growth promoter and/or growth promoter-inhibitor ratio. Although this study measured extractable concentrations of only 2 growth regulators in entire buds, rather than in specific tissues, and tissue sensitivity to the growth regulators was not determined, flower bud break of 'Climax' rabbiteye blueberry appears to be associated with the relative balance of endogenous growth inhibitors and growth promoters.

LITERATURE CITED

Alvim, R., S. Thomas, and P. F. Saunders. 1978. Seasonal variation in the hormone control of willow. II. Effect of photoperiod on growth and abscisic acid content of trees under field conditions. Plant Physiol. 62:779-780.

Austin, M. E., B. C. Mullinix, and J. S. Mason. 1982. Influence of chilling on growth and flowering of rabbiteye blueberries. HortScience 17:768-769.

Bachelard, E. P. and F. Wightman. 1974. Biochemical and physiological studies on dormancy release in tree buds. III. Changes in endogenous growth substances and a possible mechanism of dormancy release in overwintering vegetative buds of *Populus balsamifera*. Can. J. Bot. 52:1483-1489.

Bennett, J. P. and F. Skoog. 1938. Preliminary experiments on the relation of growth-promoting substances to the rest period in fruit trees. Plant physiol. 13:219-225.

Corgan, J. N. and G. C. Martin. 1971. Abscisic acid levels in peach floral cups. HortScience 6:405-406.

Darrow, G. M. 1942. Rest period requirements for blueberries. Proc. Amer. Soc. Hort. Sci. 41:189-194.

Engelbrecht, L. 1971. Cytokinins in buds and leaves during growth, maturity and aging (with a comparison of two bio-assays). Biochem. Physiol. Pflanz. 162: 547-558.

Gilreath, P. R. and D. W. Buchanan. 1981. Temperature and cultivar influences on the chilling period of rabbiteye blueberry. J. Amer. Soc. Hort. Sci. 106:625-628.

Mielke, E. A. and F. G. Dennis, Jr. 1975. Hormonal control of flower bud dormancy in sour cherry *(Prunus ceraus L.)*. II. Levels of abscisic acid and its water soluble complex. J. Amer. Soc. Hort. Sci. 100:287-290.

Nitsch, J. P. 1957. Photoperiodism in woody plants. Proc. Amer. Soc. Hort. Sci. 70:526-544.

Nooden, L. D. and J. A. Weber. 1978. Environmental and hormonal control of dormancy in terminal buds of plants, pp. 221-268. In: M. E. Clutter (ed.) Dormancy and development arrest. Academic Press, New York.

Ramsay, J. and G. C. Martin. 1970a. Seasonal changes in growth promoters and inhibitors in buds of apricot. J. Amer. Soc. Hort. Sci. 95:569-574.

Ramsay J. and G. C. Martin. 1970b. Isolation and identification of a growth inhibitor in spur buds of apricot. J. Amer. Soc. Hort. Sci. 95:574-577.

Richardson, E. A., S. D. Seeley, and D. R. Walker. 1974. A model for estimating the completion of rest for 'Redhaven' and 'Elberta' peach trees. HortScience 9:331-332.

Seeley S. 1990. Hormonal transduction of environmental stresses. HortScience 25:1369-1376.

Seeley, S. D. and L. E. Powell, Jr. 1981. Seasonal changes of free and hydrolyzable abscisic acid in vegetative apple buds. J. Amer. Soc. Hort. Sci. 106:405-409.

Spiers, J. M. 1976. Chilling regimes affect bud break in 'Tifblue' rabbiteye blueberry. J. Amer. Soc. Hort. Sci. 101:88-90.

Spiers, J. M. and A. D. Draper. 1974. Effect of chilling on bud break in rabbiteye blueberry. J. Amer. Soc. Hort. Sci. 99:398-399.

Trewavas, A. J. 1982. Growth substance sensitivity: the limiting factor in plant development. Physiol. Plant. 55:60-72.

Wareing, P. F. and P. F. Saunders. 1971. Hormones and dormancy. Annu. Rev. Plant Physiol. 22:261-288.

Weiler, E. W. 1980. Radioimmunoassays for transzeatin and related cytokinins. Planta 149:155-162.

Weiler, E. W. 1982. An enzyme-immunoassay for cis(+)-abscisic acid. Physiol. Plant. 54:510-514.

Weiler, E. W. 1984. Immunoassay of plant growth regulators. Ann. Rev. Plant Physiol. 35:85-95.

Weiler, E. W., P. S. Jourdan, and W. Conrad. 1981. Levels of indole-3-acetic acid in intact and decapitated coleoptiles as determined by a specific and highly sensitive solid-phase enzyme immunoassay. Planta 153:561-571.

Wood, B. W. 1983. Changes in indoleacetic acid, abscisic acid, gibberellins and cytokins during bud break in pecan. J. Amer. Soc. Hort. Sci. 108:333-338.

Wright, S. T. C. 1975. Seasonal changes in the levels of free and bound abscisic acid in blackcurrant *(Ribes nigrum)* buds and beech *(Fugus sylvaticus)* buds. J. Expt. Bot. 26:161-174.

Growth Responses of Southern Highbush and Rabbiteye Blueberry Cultivars at Three Southern Locations

D. J. Makus
J. M. Spiers
K. D. Patten
E. W. Neuendorff

SUMMARY. Southern highbush blueberry cultivars Georgiagem and O'Neal and rabbiteye blueberry cultivars Climax and Tifblue were grown at three southern locations for six years. There were differences in soil nutrient and physical properties between the locations. Rabbiteye plants were larger in root and shoot biomass, plant volume and fruit yield than were southern highbush plants at the termination of the evaluation period. Performance of cultivars within blueberry type (species) was similar. Rabbiteye blueberry plants performed better than the southern highbush plants at the Overton, TX

D. J. Makus is Research Horticulturist, U.S. Department of Agriculture, Agricultural Research Service, Booneville, AR 72972.

J. M. Spiers is Research Horticulturist, U.S. Department of Agriculture, Agricultural Research Service, Poplarville, MS 39470.

K. D. Patten is Associate Professor, Texas A&M University Agricultural Research & Extension Center, Overton, TX 75684.

E. W. Neuendorff was Research Associate, Texas A&M University Agricultural Research & Extension Center, Overton, TX 75684.

This paper is dedicated to the memory of Elizabeth W. Neuendorff, born August 13, 1957; deceased November 24, 1994.

[Haworth co-indexing entry note]: "Growth Responses of Southern Highbush and Rabbiteye Blueberry Cultivars at Three Southern Locations." Makus, D. J. et al. Co-published simultaneously in *Journal of Small Fruit & Viticulture* (Food Products Press, an imprint of The Haworth Press, Inc.) Vol. 3, No. 2/3, 1995, pp. 73-82; and: *Blueberries: A Century of Research* (ed: Robert E. Gough, and Ronald F. Korcak) Food Products Press, an imprint of The Haworth Press, Inc., 1995, pp. 73-82. Single or multiple copies of this article are available from The Haworth Document Delivery Service [1-800-342-9678, 9:00 a.m. - 5:00 p.m. (EST)].

and Poplarville, MS locations. Growth and yield characteristics of both blueberry types at the Booneville, AR location were similar and rabbiteye blueberry plant performance poorer than at the more southern locations. *[Article copies available from The Haworth Document Delivery Service: 1-800-342-9678.]*

INTRODUCTION

Southern highbush (predominantly *V. corymbosum* L.) are very similar to highbush cultivars (cvs) in fruit quality, but require fewer chilling hours than the latter. Rabbiteye cultivars (*V. ashei* Reade) flower about the same time or earlier in the spring, but mature fruit several weeks later than southern highbush. Chilling requirements for 'Climax,' 'Georgiagem,' and 'Tifblue' have been reported as 450, 400, and 650 hrs below 45°F (7.2°C), respectively (Austin et al., 1982; Austin and Bondari, 1987).

The potential to grow a low chill requiring, less frost susceptible, early and high-quality fruiting blueberry species in the South has generated much interest. Although earliness and fruit quality are desirable attributes, southern highbush need to be extensively evaluated in southern climates with respect to adaptation and long-term performance before large scale production is attempted.

In fall of 1989, a 6-year-old regional trial involving seven locations was terminated. At three locations, cultivars of 2 southern highbush and 2 rabbiteye blueberries, common to all three locations, were analyzed for total plant biomass (stems and roots) and the previous season's yield. Available soil nutrients and some physical soil properties were also documented at each site.

Our objective was to compare the plant biomass performance of these blueberry types over a range of edaphic and climatic conditions in the South where commercial production of these blueberry types might be feasible.

MATERIALS AND METHODS

In spring of 1984, two southern highbush cvs, Georgiagem and O'Neal, and two rabbiteye cvs, Climax and Tifblue, were planted in Booneville, AR, Overton, TX, and Poplarville, MS using the establishment practices recommended at each location. Soil types for these sites were a Leadvale silt loam (fine-silty, siliceous Fragiudult), Bowie fine sandy loam (fine-loamy, siliceous, thermic Plinthic Paleudult), Ruston fine sandy loam (fine-loamy, siliceous, thermic Typic Paleudult), respectively. The locations were at 35°06', 32°22', and 30°40' latitude, respectively.

At Booneville and Overton, plants were grown on 20.3 cm (8 in) raised beds. Peatmoss was incorporated into the planting hole and plants were mulched with sawdust for the duration of the experiment. Ammonium nitrate and ammonium sulfate were the primary sources of N-fertilizer used at each site, respectively. Water was applied by drip irrigation as needed.

During the 1988-1989 winter and 1989 summer, soil was sampled at 15 cm (6 in), 30 cm (12 in), and 45 cm (18 in) depth and analyzed for pH, organic matter (OM), electrolytes (E.C.), P, K, Ca, Mg, Al, Mn, Fe, S, Zn, and Cu (Plank, 1982). In late fall of 1989, prior to plant destruction, plant volume (height × width × depth) were determined. Plants were removed from the soil with a soil mass 1.5 ft (0.5 m) in radius from the crown of the plant. Roots growing 0 to 15 cm (0 to 6 in) and 15 to 30 cm (6 to 12 in) from the soil surface were removed, washed and force-air dried at 70°C (158°F). The portion of the plant above the soil (with leaves removed) was similarly dried and weighed.

Data were analyzed using PROC GLM of SAS (Version 6.04). Locations were treated as main-plots and soil depth or blueberry type (southern highbush and rabbiteye) as sub-plots. The three replications were regarded as completely random.

RESULTS

There were differences between locations in all soil attributes measured except Fe and Zn (Table 1). Zinc levels were not influenced by location or soil depth (data not shown). The AR location had the highest organic matter, and Ca and Cu soil test levels. The MS site had the lowest pH and highest P, K, Mg, Al, Mn, and S levels.

All soil attributes decreased with soil depth except Mg and S, which increased, and Ca, which did not change with soil depth.

TABLE 1. Soil analysis at three locations, from samples taken at three depths. Data based on pooled winter 1988-1989 and summer 1989 soil sample analyses.^z

	pH	OM (%)	E.C. (dS/M)	P	K	Ca	Mg	Al	Mn	Fe	S	Cu
								ppm				
Location:												
Bnvl, AR	5.6 a	3.4 a	0.09 b	8 c	73 b	1207 a	84 b	143 b	78 b	63 a	43 b	0.53 a
Overton, TX	5.7 a	0.8 c	0.08 b	21 b	57 c	318 c	65 c	64 c	4 c	71 a	16 c	0.27 b
Popvl, MS	5.0 b	2.3 b	0.16 a	29 a	148 a	548 b	104 a	232 a	141 a	62 a	55 a	0.33 b
	**y	**	**	**	**	**	**	**	**	NS	**	**
Soil Depth:												
15 cm	5.3 b	2.6 a	0.13 a	32 a	120 a	767 a	77 b	131 a	96 a	86 a	31 b	0.49 a
30 cm	5.6 a	2.1 b	0.11 b	17 b	88 b	667 a	78 b	160 a	81 b	67 ab	35 b	0.38 b
45 cm	5.5 ab	1.9 b	0.09 b	9 c	72 c	641 a	98 a	148 a	47 c	42 b	47 a	0.26 c
	**	**	**	**	**	NS	**	**	**	*	**	**
Polynomial fit:												
Bnvl, AR	—	L**	L**	L**	L.07	L*	L*,Q*	L*	L**	—	Q**	L**
Overton, TX	—	—	—	L**	L*	—	—	—	—	—	—	L*
Popvl, MS	L**	L**	L*	L**	L**,Q*	—	L**	Q**	L**	L**	L**,Q*	L**

^z All location × depth interactions were significant (P < 0.05). See polynomial fits in above table.
^y **, *, NS = significant at P = 0.01, P = 0.05, and not significant, respectively. The Ryan-Einot-Gadriel-Welsch multiple 'F' test was used to determine mean separation at the significance levels shown in Table 1.

76

These gradients were characteristic of the AR and MS sites. The only soil nutrient gradients occurring at the TX site were for P, K, and Cu.

Plant root, shoot (stem), volume, fruit yield, and root to shoot ratio were lowest in plants grown at the AR location (Table 2). Both the MS and TX locations were similar in plant biomass production. Because cultivars were consistent within type (species), they are presented by type for simplification. Southern highbush cultivars were smaller in root, shoot, volume, and yield attributes compared to rabbiteye cultivars. The root to shoot ratio was higher in southern highbush cultivars. Southern highbush cultivars had a higher percentage of their roots in the 0 to 15 cm (0 to 6 in) sampling zone than did rabbiteye cultivars ($P < 0.07$).

At the AR location, both blueberry types performed similarly. Rabbiteye cultivars had greater biomass attributes, compared to southern highbush cultivars, at the MS and TX locations (Figure 1).

Yield and shoot weight, total root weight, and plant volume were correlated in both blueberry types (Table 3). Plant volume was correlated with shoot weight and total root weight. Soil bulk density at the 0 to 15 cm (0 to 6 in) depth was correlated with the root to stem ratio. These simple linear correlations by blueberry type were higher in rabbiteye than in southern highbush cultivars. Bulk density at the 15 to 30 cm (6 to 12 in) depth was correlated ($r = 0.467$, $P = 0.05$) with root weight at the 15 to 30 cm (6 to 12 in) depth in southern highbush only. No additional correlations were observed between soil bulk density and the attributes measured in Table 2.

DISCUSSION

Soil bulk densities for the 0 to 15 cm (0 to 6 in) and 15 to 30 cm (6 to 12 in) depths at the Booneville, Overton, and Poplarville sites were 1.35/1.47, 1.43/1.46, and 1.36/1.49, respectively. Statistical comparisons were not possible in this study, but there did not appear to be major differences between sites in bulk density. As expected, bulk density increased with soil depth. The high organic matter content of the AR soil is not characteristic of mineral soils found in that region, but probably accumulated from years of con-

TABLE 2. Effect of location on growth responses[z] of southern highbush and rabbiteye blueberry cultivars in 1989.

Main Effect	Root wt. by depth (cm)			Root wt. (% by depth)		Shoot wt.	Root Shoot Ratio	Plant Volume (m³)	Season Fruit Yield
	0-15	15-30	0-30	0-15	15-30				
Location (Loc):									
Booneville, AR	285 b	54 b	341 b	84.4 a	15.6 b	975 a	0.35 b	0.47 b	443 b
Overton, TX	728 a	158 a	886 a	84.6 a	15.4 b	1026 a	1.24 a	1.72 a	2013 a
Poplarville, MS	651 a	218 a	869 a	74.4 b	25.6 a	1323 a	0.96 a	—	2626 a
	*y	**	**	0.11	0.11	NS	0.06	**	*
Type:									
Southern HB	286 b	55 b	341 b	83.5 a	16.5 a	493 b	1.08 a	0.42 b	474 b
Rabbiteye	824 a	233 a	1056 a	78.8 a	21.2 a	1724 a	0.62 b	1.78 a	2914 a
	**	**	**	0.07	0.07	**	**	**	**
Interaction:									
Loc X Type	**	*	**	NS	NS	**	0.08	**	**

[z] All weights are in g. Root and shoots are dry weight basis. Fruits are fresh weight basis.
[y] *, **, NS = significant at P = 0.05, P = 0.01 and not significant, respectively. Prob. > 'F' value given in cases where means are declared significantly different by the Ryan-Einot-Gabriel-Welch multiple 'F' test (∝ = 0.05) but where P was > 0.05 as determined by LSMEANS of PROC GLM in SAS.

FIGURE 1. Plant shoot weight, root weight, volume, and fruit yield interactions between location and blueberry type. Bars with different letters are significantly different (P < 0.01). Plant volume measurements were not taken at the MS location.

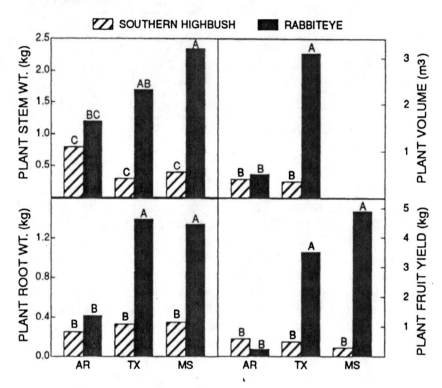

tinuous use of sawdust mulch and the initial use of peatmoss for blueberry plant establishment.

Soil texture may effect root development and subsequent plant size. Haby et al. (1986) observed that when 'Tifblue' rabbiteye plants were grown in clay, sandy loam and loamy sand soils mixed with peatmoss (1:1), top:root ratios increased, respectively, and plant weight gain was positively correlated with the percentage of sand in the soil. Rabbiteye and (high chill-requiring) highbush blueberry plants (averaged across cultivars in each group) produced similar yields (per acre) when grown in a Linker fine sandy loam soil in Bald Knob, AR (Moore et al., 1984). In that study, 'Climax'

TABLE 3. Selected linear correlations (R^2) common to both blueberry types of several biomass attributes and soil bulk density (BD).

	Southern Rabbiteye	Highbush
Yield vs. Stem wt.	0.637**[z]	0.580**
Yield vs. Root[y] wt.	0.640**	0.554*
Yield vs. Plant vol.	0.798**	0.632*
Plant vol. vs. Stem wt.	0.679**	0.624*
Plant vol. vs. Root[y] wt.	0.879***	0.733**
BD (0 to 15 cm) vs.		
Root:Stem ratio	0.535*	0.459*

[z] *, **, *** = $P < 0.05$, $P < 0.01$, and $P < 0.001$, respectively.
[y] Total root weight.

produced less fruit than 'Tifblue' or any of the highbush cultivars grown. At Bald Knob, 'Climax' suffered greater fruit bud mortality than 'Tifblue' (69 vs. 1%) during a January 20, 1985 low temperature of $-13°F$ ($-25°C$) (Clark et al., 1985). In the 1984 southern regional germplasm evaluation plantings at Clarksville, AR and Overton, TX, southern highbush and highbush cultivars had less winter freeze and spring frost damage than corresponding rabbiteye cultivars during the 1988-89 winter (Patten et al., 1991). Bud hardiness in addition to soil texture would appear to limit production of some rabbiteye cultivers in the upper mid-south.

Other edaphic conditions which could also affect root, and then subsequent shoot development, are soil temperature, water potential and gas diffusion. Warmer soil temperatures may modulate root growth during the dormant period. Highbush blueberry roots are capable of growing year-round (Abbott and Gough, 1987a). Long-term flooding can be deleterious to both rabbiteye (Crane and Davies, 1988) and highbush blueberry plants (Abbott and Gough, 1987b). The Booneville site was susceptible to a water table approaching the soil surface at the base of the raised bed during mid-winter for durations as long as 6 weeks.

Cultural management may also influence blueberry root biomass. Sites having the largest root systems had soil testing higher in P and lower in Ca. Rabbiteye blueberry plants fertilized with calcium nitrate or ammonium sulfate were smaller than controls or plants given NH_4NO_3 as fertilizers (Spiers, 1987.) In this experiment, soil Al level was not a factor in root growth; however, increasing extractable Al, by amending the soil pH by the addition of $Al_2(SO_4)_3$, resulted in growth reduction in *V. ashei* cultivers grown at three locations (Peterson et al., 1987). Nitrogen form and rate has been reported to effect highbush blueberry root mass, but negative results have also been documented (Eck, 1988). Similar contradictions exist regarding the effect of exchangeable soil Ca levels on blueberry plant growth (Eck, 1988). Incorporation of organic matter, irrigation, and mulching have been reported to influence 'Tifblue' root mass and distribution (Spiers, 1986).

CONCLUSION

Five years after establishment, rabbiteye plants had 3 times the root mass, 3.5 times the shoot mass, 4 times the plant volume and 6 fold higher yield than did southern highbush plants. Although southern highbush had fewer roots (by weight) than did rabbiteye plants, they had a higher percentage of their roots in the upper 0 to 15 cm (0 to 6 in) of the soil profile compared to rabbiteye plants. Cultivars were consistent within blueberry type for the biomass attributes measured. Plants of both types performed better in southern MS and eastern TX, than those growing in western AR.

GROWER BENEFITS

The rabbiteye cultivers Climax and Tifblue produced larger shoots and roots and had greater fruit yields than did the southern highbush cultivers Georgiagem and O'Neal grown in two of three southern locations. Both blueberry types produced larger plants on lighter-textured soils.

LITERATURE CITED

Abbott, J.D. and R.E. Gough. 1987a. Seasonal development of highbush blueberry roots under sawdust mulch. J. Amer. Soc. Hort. Sc. 112:60-62.

Abbott, J.D. and R.E. Gough. 1987b. Growth and survival of the highbush blueberry in response to root zone flooding. J. Amer. Soc. Hort. Sci. 112:603-608.

Austin, M.E. et al. 1982. Influence of chilling on growth and flowering of rabbit-eye blueberries. HortScience 17:768-769.

Austin, M.E. and K. Bondari. 1987. Chilling hour requirement for flower bud expansion of two rabbiteye and one highbush blueberry shoots. HortScience 22: 1247-1248.

Clark, J.R. et al. 1986. Cold damage to flower buds of rabbiteye blueberry culti-vars. Ark. Farm Res., November-December.

Crane J.H. and F.S. Davies. 1988. Flooding duration and seasonal effects on growth and development of young rabbiteye blueberry plants. J. Amer. Soc. Hort. Sci. 113:180-184.

Eck, P. 1988. Blueberry Science. Rutgers Uni. Press, New Brunswick, NJ.

Haby, V.A. et al. 1986. Response of container-grown rabbiteye blueberry plants to irrigation water quality and soil type. J. Amer. Soc. Hort. Sci. 111:332-337.

Moore, J.N. et al. 1984. Blueberry cultivar performance in a transition zone. Ark. Farm Res., July-August.

Patten et al. 1991. Cold injury of southern blueberries as a function of germplasm and season of flower bud development. HortScience 26:18-20.

Peterson D.V. et al. 1987. Effects of soil-applied elemental sulfur, aluminum sulfate, and sawdust on growth of rabbiteye blueberries. J. Amer. Soc. Hort. Sci. 112:612-616.

Plank. C.O. 1992. Plant analysis reference procedures for the southern region of the United States. So. Coop. Ser. Bul. 368. Ga. Agri. Exp. Sta. Athens, GA.

Spiers, J.M. 1987. Effect of fertilization rates and sources on rabbiteye blueberry. J. Amer. Soc. Hort. Sci. 112:600-603.

Spiers, J.M. 1986. Root distribution of 'Tifblue' rabbiteye blueberry as influenced by irrigation, incorporated peatmoss, and mulch. J. Amer. Soc. Hort. Sci. 111: 877-880.

SESSION III:
BLUEBERRY GENETICS CONTINUED–
BLUEBERRY DISEASES
Moderator: Jim Hancock

Characterization and Detection
of Blueberry Scorch Carlavirus
and Red Ringspot Caulimovirus

Bradley I. Hillman
Diane M. Lawrence
Barbara T. Halpern

SUMMARY. Detection systems for blueberry scorch virus (BBScV), a member of the carlavirus group of RNA-containing plant viruses, and red ringspot virus (BBRRV), a member of the caulimovirus group of DNA-containing plant viruses, have been developed. The most sensitive method for detection of both viruses was polymerase chain

Bradley I. Hillman, Diane M. Lawrence, and Barbara T. Halpern, Department of Plant Pathology, Rutgers University, New Brunswick, NJ 08903.

[Haworth co-indexing entry note]: "Characterization and Detection of Blueberry Scorch Carlavirus and Red Ringspot Caulimovirus." Hillman, Bradley I., Diane M. Lawrence, and Barbara T. Halpern. Co-published simultaneously in *Journal of Small Fruit & Viticulture* (Food Products Press, an imprint of The Haworth Press, Inc.) Vol. 3, No. 2/3, 1995, pp. 83-93; and: *Blueberries: A Century of Research* (ed: Robert E. Gough, and Ronald F. Korcak) Food Products Press, an imprint of The Haworth Press, Inc., 1995, pp. 83-93. Single or multiple copies of this article are available from The Haworth Document Delivery Service [1-800-342-9678, 9:00 a.m. - 5:00 p.m. (EST)].

reaction (PCR). Nucleic acid spot hybridization was somewhat less effective. Transmission experiments confirmed that at least one aphid species is capable of transmitting BBScV. *[Article copies available from The Haworth Document Delivery Service: 1-800-342-9678.]*

KEYWORDS: Blueberry; Virus; Nucleic acid; Detection; Polymerase chain reaction; Transmission

INTRODUCTION

Viruses cause several diseases in blueberry (*Vaccinium corymbosum* L). Two viruses of importance to blueberry growers in different regions of North America are blueberry scorch (BBScV) and red ringspot virus (BBRRV). BBScV is associated with similar blighting diseases on the east and west coasts of North America. On the east coast, the disease associated with BBScV is referred to as Sheep Pen Hill Disease, named after the geographic focus of the disease in New Jersey (Stretch, 1983; Podleckis et al., 1986; Cavileer et al., 1994; Martin et al., 1992). On the west coast, the disease is called blueberry scorch, referring to the most pronounced symptoms associated with infection by BBScV (Martin and Bristow, 1988). These diseases are similar in symptomatology, and at least three different strains of BBScV have been identified (Cavileer et al., 1994). BBScV is a filamentous virus with an RNA genome of 8.5 kilobases (kb) (Martin and Bristow, 1988; Cavileer et al., 1994). Many functions of the six genes on the BBScV genome are known or suspected, and these are summarized in Figure 1A. This information has been gained through sequence analysis of complementary DNA clones representing all of the BBScV genome, and a full-length clone from which infectious RNA can be transcribed (Cavileer et al., 1994; Lawrence and Hillman, unpublished).

Red ringspot describes leaf and stem symptoms of a disease of blueberry and cranberry caused by BBRRV, an icosahedral DNA-containing virus (Kim et al., 1981). BBRRV is considerably less thoroughly characterized than BBScV, and has only been partially characterized at the level of primary nucleotide sequence (unpublished). A diagram based on a well-characterized caulimovirus such as cauliflower mosaic virus (CaMV) (Bonneville et al., 1988) is shown in Figure 2A with a diagram of the homologous region of the partially characterized BBRRV genome.

Both BBScV and BBRRV cause diseases that are of some economic importance, so early identification and removal of infected plants is critical to their control. Although antiserum based detection systems have been used with moderate success for their detection in the past, interference due to compounds in blueberry plants have rendered such tests somewhat unreliable. Therefore, more sensitive and reliable detection methods were sought.

The polymerase chain reaction (PCR) is a technique for amplifying to detectable levels specific nucleotide sequences that would otherwise be undetectable. PCR has gained popularity because of its speed and sensitivity. It is a particularly useful technique when sample number is relatively small. We examined the feasibility of using PCR and spot hybridization of nucleic acids for detection of BBScV and BBRRV.

Different viruses vary in their economic importance in a particular region, often depending on whether or not a given virus has been introduced to a particular region, and often depending on presence or absence of an invertebrate vector capable of transmitting the virus. R. R. Martin (Agriculture Canada, unpublished) determined that the blueberry aphid, *Fimbriaphis fimbriata* Richards, is capable of transmitting BBScV from infected to uninfected blueberry plants, but little else is known about transmission of these two viruses. We wished to determine whether a vector capable of transmitting BBScV was present on the east coast.

MATERIALS AND METHODS

Infected blueberry plants were collected from fields near Pemberton, NJ. Healthy control plants were maintained in the greenhouse. *Chenopodium quinoa*, an alternate herbaceous host for BBScV (Cavileer et al., 1994), was infected by rub-inoculation and maintained in the greenhouse. We have no alternate host for BBRRV.

Aphids were collected by Sridhar Polavarapu (Rutgers Blueberry & Cranberry Research Center) from blueberry plants in Burlington Co., NJ. Aphids were propagated in the greenhouse on uninfected blueberry plants for use in *C. quinoa* transmission experiments.

Nucleic Acid Extraction

For PCR experiments, approximately 0.1 g of plant tissue was ground to a fine powder under liquid nitrogen. Four hundred µl extraction buffer (1 M Tris-HCl, pH 8.0, 100 mM LiCl, 10 mM EDTA, 1% SDS) was added, mixed, and emulsified with 400 µl neutral phenol. Samples were incubated 5 min at 65°C, centrifuged 2 min in an Eppendorf microfuge at 14,000 RPM, and the resulting supernatant was extracted twice with chloroform. Nucleic acid was precipitated for 10 min at −70°C with 2.5 vol. EtOH, collected by centrifugation for 5 min at 14,000 RPM. Pellets were washed in 70% EtOH, dried, and resuspended in 20 µl water for use in RT-PCR. For spot-hybridization experiments, nucleic acid was extracted in 100 mM Tris-HCl, pH 8.0, 200 mM NaCl, 10 mM EDTA, 2% SDS. Samples were not heated before extracting. Other steps were as described above.

Probe Production

For spot hybridization experiments, cDNA clone pBBScV44, representing 3.5 kb of the BBScV genome (Cavileer et al., 1994), was randomly labelled with ^{32}P as described previously (Hillman et al., 1992).

For detection of BBScV, oligonucleotides corresponding to nucleotide residues 6,025-6040 and 6520-6535 of the BBScV genome (Cavileer et al., 1994) were used to prime amplification of a 511 base pair (bp) fragment of DNA (Figure 1A).

The sequences of the oligonucleotides are:

Oligo BBScV 11:5' - AATTCGAGCGTCAGTC - 3'
Oligo BBScV 14:5' - AAAAGTCTGGCCGCGC - 3'

A clone representing part of the BBRRV genome was kindly provided by J. M. Gillette and D. C. Ramsdell. This clone was sequenced and found to represent a portion of the BBRRV coat protein gene (unpublished). Two oligonucleotides capable of priming a 903 bp PCR product were synthesized. The sequences of the oligonucleotides are:

Oligo BBRRV 5:5' - AGGAGGTATAGCAAAG - 3'
Oligo BBRRV 6:5' - AACTGTGATGTATTGGT - 3'

PCR and RT-PCR

Randomly primed reverse transcription reactions were performed on 2-5 μl of the total nucleic acid preparation as described previously (Cavileer et al., 1994) for detection of BBScV. Polymerase chain reactions were performed on 5 μl of the RT reactions using reagents and conditions recommended by the manufacturer (Perkin Elmer). Thirty-five PCR cycles (1.5 min. at 92°C, 1.5 min at 52°C, 1 min at 72°C) were performed Techne Model 2 thermocycler. The same procedure, without the RT step which is unnecessary for DNA-containing viruses, was used for amplification of BBRRV DNA.

Nucleic Acid Spot-Hybridization

Spot-hybridization experiments were performed essentially as described previously (Hillman et al., 1992). Nucleic acid from infected tissue was prepared as above, spotted to nylon filters, and probed with ^{32}P-labeled DNA representing BBScV or BBRRV as described above.

Immunoblotting

For aphid transmission experiments, leaves of *C. quinoa* plants were homogenized in standard phosphate-buffered saline (PBS) and spotted to nitrocellulose membranes. Membranes were processed as described by Lin et al. (1990) using antiserum raised in mice against purified BBScV virions (kindly provided by Dr. E. V. Podleckis).

RESULTS AND DISCUSSION

Both spot hybridization and RT-PCR were useful for detection of BBScV. In PCR experiments, a 511 bp product was predicted based on the nucleotide sequence of BBScV (Figure 1A). A product of that size was consistently detected from purified virus, from infected *C. quinoa*, or from infected blueberry tissue (Figure 1B). The primers were capable of amplifying a fragment from infected plant material from the east or west coast. Spot-hybridization using ^{32}P-labeled cDNA clones was also useful for BBScV detection (Figure 2). The advantage of spot-hybridization was that more samples could be processed in a shorter time than with PCR. The disadvantages included use of ^{32}P in the procedure and the somewhat lower sensitivity than PCR. With PCR, we could detect femtogram levels of purified virus, and could detect a single infected flower in a group of ten uninfected flowers (unpublished). There was somewhat higher background associated with spot-hybridization, with some non-symptomatic samples that tested negative by PCR and by infectivity assays yielding light gray spots (e.g., Figure 2C-D, lane 6).

Although we have not completed the nucleotide sequence of the BBRRV genome, comparison of the sequence of clone pBBRRV-2 to other caulimovirus sequences suggests its position as indicated in Figure 3A. A PCR product of the predicted 903 bp size was amplified from the control clone or from virus preparations (Figure 3B).

To begin to examine viral genes required for invertebrate transmission of BBScV, we have performed a limited number of transmission experiments using *C. quinoa*. Groups of 20-50 aphids were transferred from uninfected blueberry plants to infected *C. quinoa* plants which were caged together with uninfected *C. quinoa*. Although the aphids did not thrive on *C. quinoa* plants, at least 1 out of every 2 uninfected plants per cage showed symptoms within 2 weeks of aphid transfer. Immunoblot experiments with BBScV antiserum confirmed that the symptomatic plants were infected (Figure 4). We have not yet investigated optimal acquisition or maximal retention times, but based on properties of other carlaviruses, we would expect short (a few seconds to a few minutes) acquisition times and short (< 1 hr) retention times, typical of non-persistent, stylet-borne viruses.

CONCLUSIONS

Considerable progress has been made over the past few years toward our understanding of the carlavirus associated with blighting diseases such as Sheep Pen Hill Disease and blueberry scorch, and the caulimovirus associated with red ringspot disease.

FIGURE 1. Genome organization and PCR detection of BBScV. A: The genome organization of BBScV was predicted from its full-length sequence. Putative functions of proteins are provided in parentheses under the corresponding genes. The relative position of the predicted 511 bp PCR product is shown below the line representing viral RNA. B: Agarose gel (1%) showing PCR products of 12 field samples (lanes 2-13), 1 ng purified viral RNA (lane 14), or two plasmid controls (lanes 15-16). Lane 1 contains the BRL 1KB ladder. Sizes of selected ladder bands are shown at the left. The gel was stained with ethidium bromide and photographed.

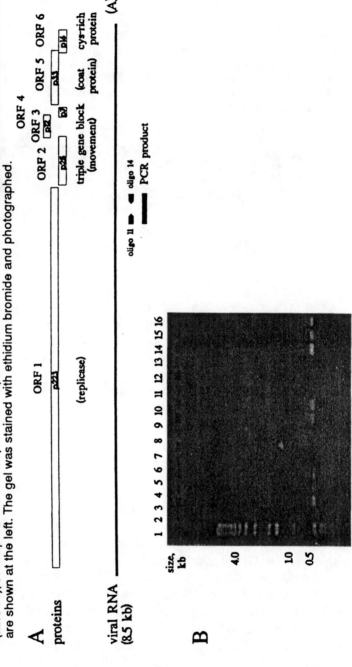

89

FIGURE 2. Detection of BBScV by nucleic acid spot-hybridization. Samples in rows A-B and C-D are two replicates of the same samples. Samples in rows A-B are: Lane 1, plasmid control (1 ng); 2, viral RNA (1 ng); 3-4, symptomless blueberries; 5, infected blueberry; 6, dogwood (non-host of BBScV); 7, uninfected *C. quinoa*; 8-11, BBScV-infected *C. quinoa*. Rows C-D: lanes 1-11 are field blueberry samples. All blueberry samples were *V. corymbosum* cv. Weymouth.

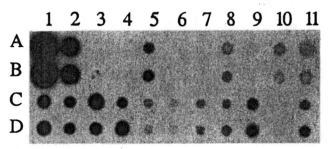

Preliminary experiments performed to examine transmission properties of BBScV confirm that aphids collected in New Jersey are capable of transmitting the virus associated with Sheep Pen Hill Disease on the east coast. It is also particularly important that we can transmit BBScV from *C. quinoa* to *C. quinoa*, since this allows us to investigate viral factors associated with aphid transmission by use of mutagenesis of our infectious full-length cDNA clones.

GROWER BENEFITS

Useful and sensitive PCR detection systems will facilitate diagnosis of the diseases discussed in this paper. We are transferring this technology from our New Brunswick, NJ lab, which is 50-80 miles from the major blueberry growing area of the state, to the Rutgers Blueberry & Cranberry Research Center in Chatsworth, close to most of the major growers in the state. The Center is now set up for PCR detection. Potential problems associated with plants suspected to be infected with BBScV can be positively diagnosed quickly, and such plants removed. Also, breeders at the center wishing to exclude BBScV and BBRRV from their stock plants will have the means to monitor for presence of these viruses. Evidence that BBScV

FIGURE 3. Partial organization and PCR detection of BBRRV. A: Clone pBBRRV-2 was positioned relative to a typical circular dsDNA caulimovirus genome based on its nucleotide sequence. The position of the predicted 908 bp PCR product is shown below the clone. B: Agarose gel (1%) showing PCR products amplified from pBBRRV-2 or viral DNA. The gel was stained and photographed as in Figure 1.

A

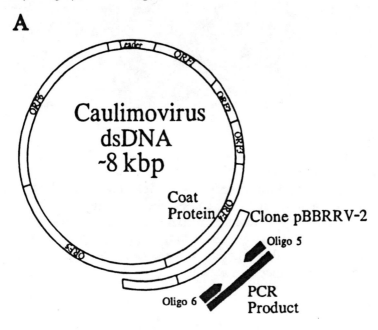

B

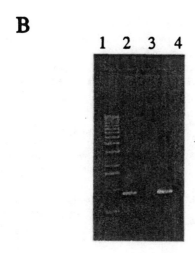

FIGURE 4. Immunoblot of *C. quinoa* plants in aphid transmission experiment. Samples from uninoculated (Lane 1), BBScV-infected (Lane 2) and two symptomatic plants exposed to aphids that were fed on a BBScV-infected plant (Lanes 3-4) were spotted to nitrocellulose and probed with anit-BBScV serum as described in the text.

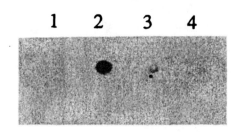

can be transmitted by aphids on the east coast provides growers with a target for control of BBScV-associated diseases.

Among the long-term benefits of this research is initial progress toward transgenic, virus-resistant blueberry plants. Although the goal of virus-resistant plants is difficult to achieve even for more tractable crops such as tomato or tobacco, our progress in understanding the molecular genetics of these two viruses makes this a realistic goal for blueberries as well.

LITERATURE CITED

Bonneville, J.M., T. Hohn, and P. Pfeiffer. 1988. Reverse transcription in the plant virus, cauliflower mosaic virus. In E. Domingo, J.J. Holland, & P. Ahlquist (Eds.), *RNA Genetics* (pp. 23-42). Boca Raton: CRC Press.

Cavileer, T.D., B.T. Halpern, D.M. Lawrence, E.V. Podleckis, R.R. Martin, and B.I. Hillman.1994. Nucleotide sequence of the carlavirus associated with blueberry scorch and similar diseases. *Journal of General Virology* 75: 711-720.

Hillman, B.I., Y. Tian, P.J.Bedker, and M.P. Brown.1992. A North American hypovirulent isolate of the chestnut blight fungus with European isolate-related dsRNA. *Journal of General Virology* 73: 681-686.

Kim, K.S., D.C. Ramsdell, J.M. Gillette, and J.P. Fulton.1981. Virions and ultrastructural changes associated with blueberry red ringspot disease. *Phytopathology* 71: 673-678.

Lin, N.S., Y.H. Hsu, and H.T. Hsu. 1990. Immunological detection of plant viruses and mycoplasmalike organism by direct tissue blotting on nitrocellulose membranes. *Phytopathology* 80: 824-828.

Martin, R.R. and P.R. Bristow. 1988. A carlavirus associated with blueberry scorch disease. *Phytopathology* 78: 1636-1640.

Martin, R.R., S.G. MacDonald, and E.V. Podleckis. 1992. Relationships between blueberry scorch and Sheep Pen Hill viruses of highbush blueberry. *Acta Horticulturae* 308: 131-139.

Podleckis, E.V., R.F. Davis, A.W. Stretch, and C.P. Schultz.1986. Flexuous rod particles associated with Sheep Pen Hill Disease of highbush blueberries. *Phytopathology* 76: 1065.

Stretch, A.W. 1983. A previously undescribed blight disease of highbush blueberries in New Jersey. *Phytopathology* 73: 375.

RELATED POSTERS

Infection of Cold-Injured Blueberry Stems by *Botryosphaeria dothidea*

W. O. Cline

SUMMARY. A blueberry field in Bladen County, NC was examined biweekly or monthly for the first 3 yr to determine conditions associated with high plant mortality in young bushes. Plants grew profusely, did not become completely dormant, and 139 of 500 bushes were cold-injured at first frost in November 1992. Cold-injured stems (ca. 10-30 cm in length) developed a characteristic dead, hook-shaped tip which persisted throughout the following growing season. In 1993, the incidence of *B. dothidea* in stems injured by cold the previous November was 19%, 39%, and 88% for March, May, and June, respectively. Widespread infection by *B. dothidea* following cold injury could account for past observations of field epidemics 1-2 yr after planting. *[Article copies available from The Haworth Document Delivery Service: 1-800-342-9678.]*

KEYWORDS: Blueberry stem blight; Cold hardiness; Plant wounding

W. O. Cline is Researcher/Extension Specialist, Department of Plant Pathology, North Carolina State University, Raleigh, NC 27695.

[Haworth co-indexing entry note]: "Infection of Cold-Injured Blueberry Stems by *Botryosphaeria dothidea*." Cline, W. O. Co-published simultaneously in *Journal of Small Fruit & Viticulture* (Food Products Press, an imprint of The Haworth Press, Inc.) Vol. 3, No. 2/3, 1995, pp. 95-98; and: *Blueberries: A Century of Research* (ed: Robert E. Gough, and Ronald F. Korcak) Food Products Press, an imprint of The Haworth Press, Inc., 1995, pp. 95-98. Single or multiple copies of this article are available from The Haworth Document Delivery Service [1-800-342-9678, 9:00 a.m. - 5:00 p.m. (EST)].

INTRODUCTION

Highbush blueberries (*Vaccinium corymbosum* L.) planted in low-lying high organic soils in southeastern NC have a history of severe stem blight caused by *Botryosphaeria dothidea* (Moug:Fr.) Ces. & de Not. (syn. *B. ribis* Gross. & Duggar). Stem blight is generally not observed in new plantings until the second and third seasons (Creswell and Milholland, 1987). Much uncertainty has existed concerning the reason for these high disease areas. *Botryo-sphaeria* species are known to colonize pruning or other wounds (Brown and Britton, 1986; Milholland, 1972; Reilly and Okie, 1984) but the wounding or other phenomenon resulting in these infections was unknown. Disease levels could not be attributed to cultivar since both resistant and susceptible cultivars had been tried in the past with high levels of infection on both. The purpose of this experiment was to determine predisposing conditions associated with observed high plant mortality in young bushes.

MATERIALS AND METHODS

A 0.2 ha planting (cv. Murphy) in Bladen County was examined biweekly or monthly for the first 3 yr. The planting was established by the grower in February of 1991 using bare-rooted hardwood cuttings set with a tractor-mounted mechanized transplanter. The field was visited regularly but given only cursory examination until fall cold injury in 1991 and subsequent stem blight infections in the warm months of 1992 prompted more detailed investigation. When cold injury again occurred in November of 1992, a count was made of the number of plants out of 500 with visibly wilting succulent stems, these stems evidencing damage by freezing temperatures. In March, May, and June, 100 cold-injured stems were collected and isolated on acidified potato-dextrose agar (aPDA). Isolates of a *Botryosphaeria* species were collected and inoculated onto a susceptible blueberry cultivar to verify pathogenicity.

RESULTS AND DISCUSSION

Plants grew profusely, did not become completely dormant, and 139 of 500 bushes were cold-injured at first frost in November

1992. Most affected plants had more than one injured stem. Cold-injured stems (ca. 5-30 cm in length) developed a characteristic dead, hook-shaped tip which persisted throughout the following growing season. The presence of *B. dothidea* was confirmed by isolation onto aPDA; Koch's postulates were completed with randomly selected isolates inoculated onto the cultivar Bluechip. In 1993, the incidence of *B. dothidea* in stems injured by cold the previous November was 19%, 39%, and 88% for March, May, and June, respectively. In March, few visible symptoms were present other than the dried, hook-shaped stem tips. On May and June sampling dates, many of the stems which tested positive for the presence of *Botryosphaeria dothidea* had also developed characteristic stem blighting (necrosis) extending beyond the cold-injured portion of the stem.

CONCLUSION

Widespread infection by *B. dothidea* following cold injury could account for past observations of field epidemics 1-2 yr after planting. While wound invasion has long been established as a primary means of infection, no high degree of correlation could be drawn between stem blight incidence and the known wounding phenomena (pruning, mechanical harvesting, insect damage) which can at times provide a point of entry. Cold injury represents a wounding phenomenon which has a very high likelihood (88%) of resulting in stem blight infection.

GROWER BENEFITS

Blueberries are produced on over 2,900 acres in southeastern NC. In 1993, the estimated value of utilized production was 13.4 million dollars. Stem blight caused by *Botryosphaeria dothidea* is the primary disease problem experienced in new plantings, and can also cause serious losses in mature fields of susceptible cultivars. In addition to causing stem blight on blueberries throughout the southeastern US, this pathogen has a wide range of other hosts. Determination of the effects of cold injury on stem blight levels points out to growers the importance of reducing fertilizer and irrigation use in late summer to avoid excessive production of succulent, freeze-susceptible growth in the fall.

LITERATURE CITED

Brown, E. A. and K. O. Britton. 1986. *Botryosphaeria* diseases of apple and peach in the southeastern United States. Plant Dis. 70:480-485.

Creswell, T. C. and R. D. Milholland. 1987. Responses of blueberry genotypes to infection by *Botryosphaeria dothidea*. Plant Dis. 71:710-713.

Milholland, R. D. 1972. Histopathology and pathogenicity of *Botryosphaeria dothidea* on blueberry stems. Phytopathology 62:654-660.

Reilly, C. C. and W. R. Okie. 1984. Susceptibility of peach tree pruning wounds to fungal gummosis as affected by time of pruning. HortScience 19:247-248.

Inventory of Pest Resistance in Blueberry Genotypes in North Carolina

S. D. Rooks
J. R. Ballington
R. D. Milholland
W. O. Cline
J. R. Meyer

SUMMARY. Blueberry breeding in North Carolina was initiated to develop cultivars resistant to diseases and insects indigenous to the southeastern United States. As a result of 55 years of blueberry breeding effort, 109 resistant genotypes have originated. These genotypes were screened for stem canker [*Botryosphaeria corticis* (Demaree & Wilcox) Arx & Muller], stem blight [*Botryosphaeria dothidea* (Mouq.:Fr.) Ces. & de Not] and/or sharp-nosed leafhopper, vector of blueberry stunt, [*Scaphytopius magdalensis* (Provancher)]. These genotypes are currently being used in the breeding program and are also available to cooperating blueberry breeding programs. *[Article copies available from The Haworth Document Delivery Service: 1-800-342-9678.]*

KEYWORDS: *Vaccinium; V. corymbosum; V. ashei; V. elliottii; V. stamineum;* Disease resistance; Insect resistance

S. D. Rooks is Research Associate, and J. R. Ballington is Professor, Department of Horticulture; R. D. Milholland is Professor and W. O. Cline is Researcher/Extension Specialists, Department of Plant Pathology; J. R. Meyer is Professor, Department of Entomology, North Carolina State University, Raleigh, NC 27695.

[Haworth co-indexing entry note]: "Inventory of Pest Resistance in Blueberry Genotypes in North Carolina." Rooks, S. D. et al. Co-published simultaneously in *Journal of Small Fruit & Viticulture* (Food Products Press, an imprint of The Haworth Press, Inc.) Vol. 3, No. 2/3, 1995, pp. 99-110; and: *Blueberries: A Century of Research* (ed: Robert E. Gough, and Ronald F. Korcak) Food Products Press, an imprint of The Haworth Press, Inc., 1995, pp. 99-110. Single or multiple copies of this article are available from The Haworth Document Delivery Service [1-800-342-9678, 9:00 a.m. - 5:00 p.m. (EST)].

INTRODUCTION

Eastern North Carolina is the major producer of cultivated high-bush blueberries in the southeastern United States, and currently ranks fifth in production in North America. Approximately 1580 hectares are established in highbush and rabbiteye blueberries in North Carolina. In a full crop year North Carolina produces over 7,500 tons of blueberries.

The survival and growth of the blueberry industry in North Carolina depends on the cooperative breeding program to supply it with improved cultivars. This is because of the number of pest problems in *Vaccinium* which are unique to the southeastern region of the United States. The North Carolina breeding program has systematically addressed breeding for pest resistance in blueberry in this region for approximately 55 years (Ballington et al., 1993).

Resistance, or at least field tolerance to fungus diseases, especially stem canker (*Botryosphaeria corticis*) and stem blight (*Botryosphaeria dothidea*) is essential in North Carolina. Resistance to additional fungal diseases such as anthracnose fruit rot [*Calletotricium gloeosporiodes* (Peny.) Peny. & Sacc.], mummy berry [*Monilina vaccinii-corymbosi* Reade (Honey)], and phomopsis twig blight (*Diaporthe vaccinii* Shear) is also needed. Blueberry stunt, a debilitating plant disorder caused by a mycoplasma-like organism, and transmitted by the sharp-nosed leafhopper (*Scaphytopius magdalensis*), is also a serious problem in commercial blueberry fields in North Carolina.

Stem Canker

The stem canker fungus mutated to six different races in North Carolina by the early 1960s and cultivars exhibited differential susceptibility to these races (Milholland, 1969). Two additional races were identified in the 1980s (Milholland, 1984; Cline and Milholland, 1988).

Stem Blight

No standard highbush cultivars have been shown to be resistant to stem blight in laboratory screenings (Milholland, 1972; Creswell and Milholland, 1987). Resistance has been found in *Vaccinium angustifolium* Aiton, *V. elliottii* Chapman, and *V. myrtilloides* Michx., but not in wild *V. corymbosum* L. (Ballington et al., 1993).

Blueberry Stunt

No genetic resistance has been found to blueberry stunt, caused by a mycoplasma-like organism. It is vectored by the sharp-nosed leafhopper (*S. magdalensis*) in North Carolina (Hoffman, 1974; Meyer and Ballington, 1990). However, nonfeeding preference resistance to the leafhopper vector occurs in *V. ashei* Reade, *V. amoenum* Ait., *V. pallidum* Ait., and *V. elliottii*, and also in *V. arboreum* Marsh. (section *Batodendron*) and *V. stamineum* L. (section *Polycodium*), (Etzel and Meyer, 1986; Meyer and Ballington, 1990).

The purpose of this report is to provide an inventory of genotypes determined to be pest and disease resistant.

MATERIALS AND METHODS

The methods of Milholland and Galletta (1969) were used in screening for stem canker resistance; Creswell and Milholland (1987), Buckley and Ballington (1987), and Cline et al. (1993) for stem blight resistance; and Meyer and Ballington (1990) for sharp-nosed leafhopper nonfeeding preference for resistance.

RESULTS AND DISCUSSION

Genotypes resistant to stem canker, stem blight and the sharp-nosed leafhopper are summarized in Table 1. Several genotypes listed have also been shown to be resistant to anthracnose or mummy berry in field tests.

Stem Canker

Of the five southern highbush cultivars recently released in North Carolina, only 'Reveille' (1990) and 'Bladen' (1991) have any stem

TABLE 1. Blueberry genotypes determined to be resistant to insects and diseases in North Carolina.

Genotype	Parentage or Species	Ploidy or Type Cross	Resistance
Angola	Weymouth × F-6	4n	stem canker races 1,2,3,4,5,6
Bladen	NC 1171 × NC SF-12-L	4n	stem canker races 1,4
Bluechip	Croatan × US 11-93	4n	stem canker races 1,4,7
Bluecrop	CU-5 × GM-37	4n	stem canker races 2,3,4,5; stem blight parent
Bounty	Murphy × G-125	4n	stem canker races 1,4
Brightwell	Tifblue × Menditoo	6n	leafhopper[z]
Briteblue	Ethel × Callaway	6n	leafhopper
Cape Fear	US 75 × Patriot	4n	stem blight
Centurion	W-4 × Callaway	6n	stem canker races 1,4; stem blight
Crabbe-4	wild *V. corymbosum*	4n	stem canker races 1,3,4,5,6,7
Croatan	Weymouth × F-6	4n	stem canker races 1,2,5,6
Earliblue	Weymouth × Stanley	4n	stem canker 2,5,6
Ethel	wild *V. ashei*	6n	leafhopper
Legacy	Elizabeth × US 75	4n	stem blight
Morrow	Angola × Adams	4n	stem canker races 1,2,4,5; fruit anthracnose
Murphy	Weymouth × F-6	4n	stem canker races 1,2,3,4,5,6,7
Premier	Tifblue × Homebell	6n	stem canker races 1,4; stem blight; leafhopper
Powderblue	Tifblue × Menditoo	6n	fruit anthracnose
Puru	E-118 × Bluecrop	4n	stem blight
Reka	E-118 × Bluecrop	4n	stem blight

Genotype	Parentage or Species	Ploidy or Type Cross	Resistance
Reveille	NC 1171 × NC SF-12-L	4n	stem canker race 7; mummy berry
Sierra	US 169 × Bluecrop	4n	stem blight
Tifblue	Ethel × Clara	6n	stem canker race 1,3,4,5
Weymouth	June × Cabot	4n	stem canker races 2,6
Woodard	Ethel × Calloway	6n	leafhopper
B-59	*V. stamineum*	2n	leafhopper
B-85	*V. stamineum*	2n	leafhopper
G-279	US 75 × Patriot	4n	stem blight
G-478	G-180 × US 75	4n	stem blight
G-496	US 158 × G-136	4n	stem blight
G-600	G-144 × FL4-76	4n	stem blight
T-141	Tifblue × Woodard	6n	leafhopper
US 681	G-362 × NJ-US 11	5n	stem blight
NC 945	E-66 × NC 683	4n	stem canker 1,4,7
NC 1066	E-30 × M-23	4n	stem blight
NC 1146	Murphy × G-125	4n	stem canker races 1,7
NC 1376	Patriot × US 74	4n	stem canker races 1,3,4; stem blight
NC 1524	US 75 × Patriot	4n	stem blight
NC 1550	Centurion × NC 911	6n	stem canker races 1,4; stem blight
NC 1560	Centurion × Southland	6n	stem blight
NC 1611	US 75 × Patriot	4n	stem blight
NC 1622	US 75 × Darrow	4n	stem blight

TABLE 1 (continued)

Genotype	Parentage or Species	Ploidy or Type Cross	Resistance
NC 1771	US 122 × US 75	4n	stem canker race 1
NC 1776	US 75 × US 117	4n	stem canker race 1
NC 1832	7-63-3a × Premier	6n	stem blight
NC 1846	G-144 × US 141	4n	stem canker race 1
NC 1849	290-1 × G-156	4n	stem canker races 1,4
NC 1852	290-1 × G-156	4n	stem canker race 4; stem blight
NC 1871	Bluechip × NC 945	4n	stem canker race 7
NC 1872	Bluechip × NC 945	4n	stem canker race 7
NC 1877	NC 763 × Premier	6n	stem blight; leafhopper
NC 1944	Tifblue × B-46	5n	leafhopper
NC 1945	Tifblue × B-46	5n	stem canker races 1,4
NC 2013	Premier × Centurion	6n	stem blight
NC 2041	Elliot × Fla 6-11	4n	stem canker races 4,7
NC 2105	Tifblue × B-46	5n	leafhopper
NC 2125	Tifblue × B-46	5n	leafhopper
NC 2389	NC 853 × B-70	5n	stem blight
NC 2600	Bounty × NC 1944	4n × 5n	stem blight; leafhopper
NC 2601	Bounty × NC 1944	4n × 5n	stem blight; leafhopper
NC 2625	Bluechip × Fla 4-15	4n	stem blight
NC 2675	Bluechip × NC 1524	4n	stem blight
NC 2690	US 124 × US 233	6n	stem blight

Genotype	Parentage or Species	Ploidy or Type Cross	Resistance
NC 2845	*V. elliottii*	2n	stem blight; leafhopper
NC 2846	*V. elliottii*	2n	stem blight; leafhopper
NC 3028	NC 945 × Morrow	4n	stem blight
NC 3030	Morrow × Bluecrop	4n	stem blight
NC 3060	NC 2125 × Murphy	5n × 4n	stem blight; leafhopper
NC 3066	NC 2125 × Murphy	5n × 4n	leafhopper
NC 3082	NC 2105 × Murphy	5n × 4n	stem blight; leafhopper
NC 3084	NC 2125 × NC 2105	5n × 5n	leafhopper
NC 3086	NC 2125 × NC 2105	5n × 5n	leafhopper
NC 3201	NC 1406 × NC 2161	4n	stem blight
NC 3208	NC 1524 × NC 1871	4n	stem canker races 1,4
NC 3209	NC 1524 × NC 1871	4n	stem canker races 1,4
NC 3210	Bounty × Blue Ridge	4n	stem canker races 1,4
NC 3211	Bounty × Blue Ridge	4n	stem canker races 1,4
NC 3212	Bounty × Blue Ridge	4n	stem canker races 1,4
NC 3213	Bounty × Blue Ridge	4n	stem canker races 1,4
NC 3214	Bounty × Blue Ridge	4n	stem canker races 1,4
NC 3215	Bounty × Blue Ridge	4n	stem canker races 1,4
NC 3216	Bounty × Premier	5n	stem canker races 1,4
NC 3217	Bounty × Premier	5n	stem canker races 1,4
NC 3218	Bounty × Premier	5n	stem canker races 1,4
NC 3219	Bounty × Premier	5n	stem canker races 1,4

TABLE 1 (continued)

Genotype	Parentage or Species	Ploidy or Type Cross	Resistance
NC 3221	Bladen × Bounty	4n	stem canker races 1,4
NC 3222	Bladen × Bounty	4n	stem canker races 1,4
NC 3223	Bladen × Bounty	4n	stem canker races 1,4
NC 3224	Bladen × Bounty	4n	stem canker races 1,4; stem blight
NC 3225	NC 2303 × NC 1637	4n	stem canker races 1,4
NC 3259	Bladen × Bounty	4n	stem canker races 1,4
NC 3270	NC 1777 × Bounty	4n	stem canker races 1,4
NC 3271	NC 1777 × Bounty	4n	stem canker races 1,4
NC 3336	*V. angustifolium*	4n	stem blight
NC 3340	*V. angustifolium*	4n	stem blight
NC 3341	*V. angustifolium*	4n	stem blight
NC 3343	*V. angustifolium*	4n	stem blight
NC 3346	*V. angustifolium*	4n	stem blight
NC 3347	*V. angustifolium*	4n	stem blight
NC 3349	*V. angustifolium*	4n	stem blight
NC 3352	*V. angustifolium*	4n	stem blight
NC 3353	*V. angustifolium*	4n	stem blight
NC 3354	*V. angustifolium*	4n	stem blight
NC 3357	*V. angustifolium*	4n	stem blight
NC 3395	*V. elliottii*	2n	stem blight; leafhopper
NC 3396	*V. elliottii*	2n	stem blight; leafhopper

Genotype	Parentage or Species	Ploidy or Type Cross	Resistance
NC 3397	*V. elliottii*	2n	stem blight; leafhopper
NC 3398	*V. elliottii*	2n	stem blight
NC 3408	NC 3033 × NC 3060	4n × (5n × 4n)	leafhopper
NC 3410	NC 3033 × NC 3060	4n × (5n × 4n)	stem blight
NC 3425	NC 2675 × NC 3060	4n × (5n × 4n)	stem blight
NC 3483	NC 1146 × NC 2673	4n × (5n × 4n)	stem blight
NC 3488	Bounty × NC 945	4n	stem blight
NC 3554	NC 1146 × NC 2673	4n × (5n × 4n)	stem blight
NC 3590	NC 2123 × O'Neal	4n	stem blight
NC 3592	NC 2128 × O'Neal	4n	stem blight
NC 3593	NC 2128 × O'Neal	4n	stem blight
NC 3600	US 399 × G-303	4n	stem blight
NC 3652	NC 2210 × NC 1877	6n	stem blight
NC 3656	NC 1572 × NC 1877	6n	stem blight
NC 3745	NC 3034 × NC 79-62-1	2n	leafhopper
NC 3746	NC 3034 × NC 79-62-1	2n	leafhopper
NC 3748	NC 3034 × NC 79-62-1	2n	leafhopper
NC 3749	NC 3034 × NC 79-62-1	2n	leafhopper
NC 3751	NC 3034 × NC 79-62-1	2n	leafhopper
NC 3753	NC 3034 × NC 79-62-1	2n	leafhopper
NC 3754	NC 3034 × NC 79-62-1	2n	leafhopper
NC SF-12-L	Ivanhoe × NC 297	4n	stem canker races 1,4

TABLE 1 (continued)

Genotype	Parentage or Species	Ploidy or Type Cross	Resistance
NC 78-8-6	V. stamineum	2n	leafhopper
NC 78-14b-1-2	V. elliottii	2n	leafhopper
NC 84-13-1	V. elliottii × V. pallidum	2n	leafhopper
NC 84-14-3	V. elliottii	2n	leafhopper
NC 84-15-1	V. elliottii	2n	leafhopper
NC 84-15-3	V. elliottii	2n	leafhopper
NC 90-3-19	V. elliottii	2n	leafhopper

ᵃHigh level of nonfeeding preference resistant to the sharp-nosed leafhopper (leafhopper nymph growth index of 2.0 or less).

canker resistance. The standard highbush cultivar 'Bounty' has not shown stem canker, and earlier releases such as 'Angola,' 'Bluechip,' 'Croatan,' 'Morrow' and 'Murphy' are resistant to certain races.

The predominant races found to be a problem in North Carolina are races 1, 4 and 7. Another virulent race 8 discovered on 'Murphy' is a concern, but it has thus far only been found in an abandoned blueberry field (Cline and Milholland 1988). Screening is being continued on elite selections which show apparent resistance after being exposed to natural infection in the field.

Stem Blight

Buckley and Ballington (1987) determined that 'Bluecrop,' 'Morrow,' and 'Murphy' appear to be prepotent highbush parents for resistance to stem blight. As mentioned earlier resistance has been found in V. angustifolium, V. elliottii, and V. myrtilloides, but not in wild V. corymbosum. Several selections involving V. elliottii have been determined to be resistant. Since these selections also have some V. angustifolium in their background, resistance may be derived either from the later species, V. elliottii, or from both species. *Vaccinium darrowi* Camp backcross derivative genotypes including 'Cape Fear,' 'Legacy,' 'Sierra,' NC 1852, NC 2675, NC 3201, G-478, and G-600 have also been determined to be resistant to stem blight in controlled screening tests (Ballington et al., 1993; Cline et al., 1993). As with Buckley's studies on standard highbush, the likely source of resistance in these genotypes is *V. angustifolium*. All the numbered selections listed above are under serious consideration for possible release as new cultivars.

Blueberry Stunt

Genes for nonfeeding preference resistance are currently being transferred to standard and southern highbush blueberry from *V. ashei*, *V. amoenum*, *V. pallidum* and *V. elliottii*. Progenies and individual selections are screened each year. Resistance genes from *V. ashei* have been successfully incorporated into BC_2 generation genotypes, and intercrosses among relatively unrelated resistant tetraploid cultivars and determine the heritability of resistance from this source. Incorporation of resistance from the other three species has not progressed as far as with *V. ashei*.

LITERATURE CITED

Ballington, J.R., S.D. Rooks, W.O. Cline, R.D. Milholland, and J.R. Meyer. 1993. Breeding blueberries for pest resistance in North Carolina. Acta Horticultureae 346:87-94.

Buckley, B. and J.R. Ballington. 1987. Screening native *Vaccinium* species for resistance to stem blight. HortScience 22:101.

Cline, W.O. and R.D., Milholland. 1988. Identification of a new race of *Botryosphaeria corticis* on highbush and rabbiteye blueberry in North Carolina. Plant Disease 72:268.

Cline, W.O., S.D. Rooks, R.D. Milholland, and J.R. Ballington. 1993. Techniques in breeding for resistance to blueberry stem blight caused by *Botryosphaeria dothidea*. Acta Horticultureae 346:107-110.

Creswell, T.C. and R.D. Milholland. 1987. Responses of blueberry genotypes to infection by *Botryosphaeria dothidea*. Plant Disease 71:710-713.

Etzel, R.W. and J.R. Meyer. 1986. Resistance in blueberries to feeding and oviposition by the sharp-nosed leafhopper *Scaphytopius magdalensis* (Provancher). J. Econ. Ent. 79:1513-1515.

Hoffman, S.T. 1974. Blueberry stunt disease in North Carolina. M.S. Thesis, North Carolina State University.

Meyer, J.R. and J.R. Ballington. 1990. Resistance of *Vaccinium* spp. to the leaf-hopper *Scaphytopius magdalensis* (Homoptera: Cicadellidae). Ann. Entomol. Soc. Am. 83:515-520.

Milholland, R.D. 1972. Histopathology and pathogenicity of *Botryosphaeria dothidea* on blueberry stems. Phytopathology 62:654-660.

Milholland, R.D. 1984. Occurrence of a new race of *Botryosphaeria corticis* on highbush and rabbiteye blueberry. Plant Disease 68:522-523.

Milholland, R.D. and G.J. Galletta. 1969. Pathogenic variation among isolates of *Botryosphaeria corticis* on blueberry. Phytopathology 59:1540-1543.

SESSION IV:
BLUEBERRY CULTURE
Moderator: Nicholi Vorsa

Blueberry Gall Midge:
A New Pest of Rabbiteye Blueberries

Paul M. Lyrene
Jerry A. Payne

SUMMARY. Natural infestations of blueberry gall midge (*Dasineura oxycoccana* Johnson) on blueberries were studied in Florida and southeastern Georgia for three flowering seasons (1992-1994). If high populations of adult midges were present when the flower buds were beginning to expand, many rabbiteye cultivars suffered 50% to 100% crop loss. In southeast Georgia, adult midge populations normally remained low through the period when flower buds were most susceptible to damage, probably because of cool winter tempera-

Paul M. Lyrene and Jerry A. Payne, Horticultural Sciences Department, University of Florida, Gainesville, FL 32611 and Southeastern Fruit and Tree Nut Research Laboratory, USDA-ARS, 111 Dunbar Road, Byron, GA 31008.

[Haworth co-indexing entry note]: "Blueberry Gall Midge: A New Pest of Rabbiteye Blueberries." Lyrene, Paul M., and Jerry A. Payne. Co-published simultaneously in *Journal of Small Fruit & Viticulture* (Food Products Press, an imprint of The Haworth Press, Inc.) Vol. 3, No. 2/3, 1995, pp. 111-124; and: *Blueberries: A Century of Research* (ed: Robert E. Gough, and Ronald F. Korcak) Food Products Press, an imprint of The Haworth Press, Inc., 1995, pp. 111-124. Single or multiple copies of this article are available from The Haworth Document Delivery Service [1-800-342-9678, 9:00 a.m. - 5:00 p.m. (EST)].

111

tures. From Gainesville south in Florida, midge populations were normally high at the time blueberry flower buds passed through the vulnerable stages from January through early April. Rabbiteye cultivars varied greatly in susceptibility of the flower buds to the midge. 'Premier' and 'Windy' were highly susceptible. 'Powderblue' and 'Brightwell' were highly resistant. Flower buds of most southern highbush cultivars were highly resistant. Sprouting blueberry vegetative buds were also infested and killed by the midge. On highly susceptible rabbiteye cultivars in Florida, this greatly delayed full foliation in the spring. In north Florida, the midges appeared to undergo multiple generations between early January and early June, and then to make no new bud infestations until the following winter. *[Article copies available from The Haworth Document Delivery Serivce: 1-800-342-9678.]*

KEYWORDS: Rabbiteye blueberry; *Vaccinium ashei* Reade; *Dasineura oxycoccana* Johnson; Insects

INTRODUCTION

The rabbiteye blueberry (*Vaccinium ashei* Reade) has been planted on about 2000 ha in southeast Georgia and 500 ha in north Florida during the past 15 years. The plants are large, vigorous, and capable of carrying heavy fruit loads. They thrive on certain soils low in organic matter where highbush blueberries (mostly *V. corymbosum* L.) grow poorly, and the berries of some cultivars are well suited to mechanical harvest. Rabbiteyes, more than highbush, maintain high fruit firmness and flavor during the hot, rainy weather common in the southeastern United States in June and July, when the berries are ripening.

On the other hand, cultivated rabbiteyes have some disadvantages compared to highbush. The berries ripen about a month later than highbush grown in the same area. Fruit set (percentage of flowers that produce ripe berries) has been low and erratic on large rabbiteye farms in Florida and southeastern Georgia, probably due to pollination problems. This has lead to the widespread use of gibberellic acid to enhance fruit set (Mainland et al., 1979; Vanderwegen and Krewer, 1991) and to efforts to increase bumblebee populations in commercial plantings.

One of the most puzzling problems with rabbiteye blueberries in the Florida peninsula has been the frequent occurrence of high levels of flower bud abortion during the winter and early spring. This has been variously attributed to lack of chilling, freeze damage, or high temperatures during flower bud development. The problem varies in severity from year to year and from field to field. It tends to be worst after mild winters, and becomes increasingly severe as one moves southward down the Florida peninsula. Young rabbiteye plantings in their second or third year often flower and fruit well, even when mature plantings of the same cultivars have severe bud loss in nearby fields. This suggests the slow build-up of populations of some pest as the plantings mature.

On 20 February 1992, rabbiteye flower buds at the University of Florida Horticultural Unit in Gainesville were found to be heavily infested with blueberry gall midge larvae (*Dasineura oxycoccana* Johnson) (Lyrene and Payne, 1992; Payne et al., 1993). Flower buds on many cultivars disintegrated within two weeks after being infested (Figure 1). The pattern of degeneration matched what had been seen for many years with rabbiteyes in the Florida peninsula. A survey of 11 rabbiteye plantings in Florida in the spring of 1992 showed gall midge infestations and severe flower bud damage at 10 locations (25% to 90% bud loss) and no midges and no damage at one farm south of the normal rabbiteye production area (Lyrene and Payne, 1992). In both 1993 and 1994, gall midge infestations were widespread and abundant in Florida, with severe commercial losses in 1993. This paper presents observations relating to blueberry gall midge damage in Florida and southeast Georgia over the past three years and discusses possible methods of reducing future damage.

GALL MIDGE BIOLOGY AND PHENOLOGY

Much information about plant parasitic gall midges is presented by Gagné (1989). Gall midges belong to the family Cecidomyiidae in the order Diptera, or two-winged flies. More than 1,200 named species of Cecidomyiidae are known from North America, many of which are parasites of one plant species or a group of closely-related species (Gagné, 1989). The adults are tiny, delicate flies, 2-3 mm long (Figure 2). All or most feeding is done by the larvae, which are usually less than 3 mm long, legless, and white, yellow, or red (Figure 3). They feed by sucking plant juices.

FIGURE 1. Flower buds of infested (left) versus uninfested (right) rabbiteye blueberry.

Gagné (1989) listed three gall midges that have been implicated in flower bud or leaf bud damage on *Vaccinium* (either blueberry or cranberry). One species, *Prodiplosis vaccinii* (Felt), has only rarely been collected on blueberry and is of no known commercial significance. It may be a polyphagous species that only facultatively infests blueberry buds. R. J. Gagné advises us in personal communication (letter of June 6, 1994) that the remaining two names, *Dasineura oxycoccana* (Johnson) and *Dasineura cyanococci* (Felt), evidently represent the same species. He will be publishing separately on this, but advises us that specimens associated with blueberry (including Florida rabbiteyes) and cranberry from many localities in the U.S. and Canada are not distinguishable.

In cranberry, *Dasineura oxycoccana* is called the cranberry tipworm (Smith, 1903; Scammell, 1917; Mahr and Kachadoorian, 1990). Mahr and Kachadoorian (1990) reported that in a Wisconsin cranberry bog, cranberry tipworms underwent five generations in 1988 and four in 1989. The first eggs were found during late May

FIGURE 2. Adult *Dasineura oxycoccana* (Johnson) depositing eggs between the scales of flower buds.

and the last during the last week in August. Larvae were extremely abundant from the last of May through the first week in July, and were present in much smaller numbers through July and August.

Adult gall midges typically live only long enough to mate and lay eggs (Gagné, 1989). The eggs hatch within a few days of deposition. Cranberry tipworm eggs hatch into tiny, clear, headless, legless maggot-like larvae, which are the damaging stage (Mahr and Kachadoorian, 1990). When fully grown, the larvae spin silken cocoons in the damaged cranberry stem tip and pupate within this cocoon. With the last generation of the year, the larvae probably all leave the plant and make cocoons in the ground where they diapause until the next year. Adult midges emerge from the cocoons. Mahr and Kachadoorian found that the length of the tipworm life cycle in Wisconsin cranberry bogs ranged from slightly over two to about 3.5 weeks depending on the time of the year. The effectiveness of sanding cranberry beds as a control for the cranberry tipworm (Mahr, 1991) indicates that tipworm larvae spend the winter

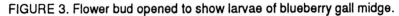

FIGURE 3. Flower bud opened to show larvae of blueberry gall midge.

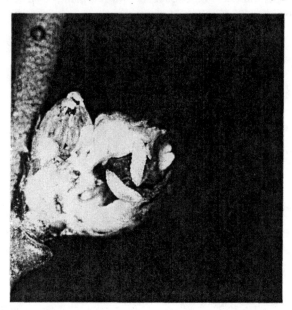

in the soil of the cranberry bog. The timing of blueberry shoot infestations in Florida and southeast Georgia indicates that the gall midge probably spends summer, fall, and the early part of the winter as larvae in the soil.

GALL MIDGES IN FLORIDA AND GEORGIA

In Florida and southeast Georgia, little damage to blueberry vegetative meristems is seen in commercial plantations after May 15, even though susceptible new growth flushes are produced by the plants throughout the summer. This indicates that adult midges are absent or rare during the summer and fall. In southeast Georgia, blueberry gall midges normally make their first appearance in late March or early April, infesting vegetative meristems during or shortly after flowering (Figures 4 and 5). In Georgia, low temperatures appear to prevent earlier emergence of the flies, allowing flower buds to pass through the vulnerable stages without being

FIGURE 4. Leaf buds infested with gall midge larvae often fail to develop fully and leaves usually do not unfold. Note larva (upper right) exiting from leaf roll.

infested. In the Florida peninsula, however, temperatures often become warm enough between late December and early April to allow adult flies to emerge at a time when rabbiteye flower buds are just beginning to expand and are highly susceptible to gall midge infestation. In Gainesville, Florida, during the winters of 1991-1992 and 1992-1993, warm temperatures in December, January, and February, resulted in periodic emergence of adult flies throughout the winter, and flower buds of many susceptible rabbiteye varieties were devastated in commercial plantings throughout north Florida. In these same years, little damage to flower buds occurred in southeast Georgia, although vegetative meristems were infested later in the spring.

The first half of the winter of 1993-1994 was colder than normal in north and central Florida, and few if any midges emerged until after flower buds on most rabbiteye cultivars in Gainesville had passed the stage most vulnerable to midges (Flower bud stages 2 and 3: Spiers, 1978). Consequently, flower buds suffered little dam-

FIGURE 5. Heavily infested leaf bud (leaf roll) showing several blueberry gall midge larvae.

age from Gainesville north, although vegetative buds became heavily infested during March. At the same time, flower buds were heavily damaged in commercial rabbiteye plantings farther south in the state. In Gainesville, almost all flower buds in a large nursery of high chill (late-flowering) rabbiteye × *V. constablaei* A. Gray hybrids were destroyed by midges during March and April. These observations indicate that warm weather during the winter predisposes rabbiteye flower buds to midge damage by allowing the flies to emerge early in the season, when the flower buds are in early bud-swell and are highly susceptible to infestation and destruction.

Several conclusions were reached regarding blueberry flower bud infestation at the Horticultural Unit in Gainesville during the three winters from 1991-1992 to 1993-1994.

1. Dormant flower buds do not become infested. Female flies apparently insert their eggs between the scales of flower buds only after the buds begin to expand. Fully dormant flower buds were never found to be infested with midge eggs or larvae. Two highly susceptible rabbiteye cultivars, 'Windy' and 'Premier,' showed the effect of bud stage on infestation in the winters of 1991-1992 and 1992-1993. Flowers of 'Windy,' an early-flowering cultivar, were destroyed before 15 February in both years, whereas the buds of 'Premier' remained dormant, healthy, and free of midges throughout February. However, as the flower buds of 'Premier' came out of dormancy in late March and early April, they were quickly infested and destroyed. On 24 February 1993, five 'Premier' branches bearing flower buds were enclosed in paper bags to exclude flies. On 13 April, after essentially all unbagged flower buds had degenerated, the bags were opened, and the branches were laden with healthy flowers near the time of anthesis.

2. The population of adult gall midges varies greatly from week to week in the same field during the winter in Gainesville, presumably due to the effect of variable weather on the emergence of adults. It is common in Gainesville during the winter to have, on the same blueberry plant at the same time, flower buds at various stages of development. Flower buds that pass through the vulnerable stage when adult gall midges are absent give rise to healthy flowers. Flower buds on the same plant that begin bud-swell when adults are abundant are destroyed. Populations of adult flies can be monitored through the winter by noting levels of infestation and destruction of flower buds of susceptible cultivars.

3. Blueberry cultivars and species vary greatly in susceptibility of the flower buds to death by gall midges. Although both flower buds and vegetative meristems can be attacked by the insect, flower-bud resistance is not necessarily associated with vegetative meristem resistance. The flower buds of most southern highbush clones at the research farm in Gainesville have shown high resistance to gall midges, but vegetative meristems of most of these are quickly killed by midge infestation.

Although we have only recognized gall midges as the cause of rabbiteye blueberry flower bud abortion in Florida for 3 flowering seasons, the same pattern of flower-bud abortion has been observed for more than 15 years. Erratic emergence of adult midges impedes evaluation of susceptibility and resistance based on field observations, but the following conclusions are supported by repeated observations. (a) Southern highbush cultivars tend to be highly resistant to flower bud damage by midges. (b) Most wild rabbiteye seedlings, including those from the southeast Georgia race of *V. ashei* as well as those from the west-Florida race, are susceptible, although level of susceptibility varies greatly from clone to clone. (c) Most rabbiteye × *V. constablaei* hybrids and backcrosses are highly susceptible. In addition to high susceptibility, these clones tend to be highly vulnerable to infestation; they flower late in the season and flower budswell begins when temperatures are warm and adult midges are abundant. (d) Among rabbiteye cultivars, 'Powderblue' and 'Brightwell' are highly resistant to flower bud damage. 'Climax,' 'Aliceblue,' 'Beckyblue,' 'Bonita,' 'Tifblue' and 'Woodard' are moderately susceptible, and 'Windy' and 'Premier' are highly susceptible. In commercial fields in Gainesville where no insecticides are used, 'Powderblue' and 'Brightwell' suffer little damage, but 'Aliceblue,' 'Beckyblue,' and 'Climax' have lost over 75% of their crop in bad years. 'Windy' and 'Premier' produce little, or no fruit in Florida if adult gall midges are abundant during budswell.

The mechanism of resistance of blueberry flower buds to gall midges is unknown. The timing of flower bud expansion was examined and discarded as an explanation. The large cultivar collection at the Horticultural Unit includes early and late flowering cultivars of both highbush and rabbiteye. Highbush flower buds repeatedly received only light to moderate midge damage, when rabbiteye flower buds at the same stage of development were repeatedly devastated. Flower bud structure was examined for susceptible and resistant clones; no obvious differences were found that could affect susceptibility to infestation. With at least one other plant parasitic gall midge, the Hessian fly, *Mayetiola destructor* (Say), susceptibility or resistance depends on the degree to which host tissues can support development of the larvae. Just as many eggs are laid on resistant as on susceptible wheat cultivars, but the larvae die or grow poorly on the resistant cultivars (Buntin et al., 1990).

Gall midge damage to blueberry vegetative meristems was studied in the field and greenhouse in Gainesville in the spring of 1994. High populations of midges were present at both sites. Blueberry taxa examined included numerous southern highbush clones from the breeding program, numerous rabbiteye clones, and several clones each of *V. darrowi* Camp, *V. elliottii* Chapman, and *V. arboreum* Marshall. Vegetative tips of *V. arboreum, V. elliottii,* and *V. ashei* were highly susceptible. All southern highbush clones examined had numerous meristems killed, but only the meristem itself, and little other tissue of the developing shoot tip was killed. Most southern highbush plants were able to foliate well despite high midge populations. On many of the rabbiteye clones, by contrast, active meristems were killed so quickly and so universally that the plants could only produce very short shoots with a few highly distorted leaves. Rabbiteye cultivars varied widely in the extent of vegetative damage. 'Climax,' one of the most susceptible, remained almost leafless late into the spring because the vegetative meristems were repeatedly killed by midges. The only taxon examined whose vegetative meristems appeared highly resistant to gall midges was *V. darrowi.* Both in the greenhouse, where pressure from the flies was extremely high, and in the field, where most rabbiteye apical meristems were killed throughout February, March, and April, few if any *V. darrowi* meristems were killed, and the plants were able to produce long, unbranched, new shoots.

OUTLOOK AND CONTROL

So far, gall midge has done little damage to blueberry flower buds in southeastern Georgia, presumably because low winter temperatures delay emergence of the flies until after rabbiteye flower buds have passed the vulnerable stage. In some years, midges heavily infest vegetative meristems of rabbiteye blueberries in southeast Georgia, reducing vegetative growth and leafiness of susceptible varieties at a critical time in berry development. On some cultivars, such as 'Climax,' midges can prevent the plant from making enough leaves to support a heavy berry crop, causing the berries to be small and low in sugar.

In north and central Florida, blueberry gall midge is the most important limiting factor in production of rabbiteye blueberries. From Gainesville south, flower buds of susceptible varieties are often ravaged, and in many years, few flower buds survive to produce flowers. Even when flower buds escape, leafing is severely curtailed on most rabbiteye varieties, reducing the size and quality of the fruit. Left uncontrolled, blueberry gall midge makes rabbit-eye blueberry production commercially infeasible south of Gainesville, Florida.

Possible methods of reducing gall midge damage include cultural methods, insecticides, and resistant cultivars.

Sanding cranberry beds has been found to reduce tipworm damage in cranberry (Barnes, 1948; Mahr, 1991). Gall midge emergence in blueberry plantings might be reduced by shallow discing beneath the blueberries at a critical time, probably in late fall or early winter, but this has not been tested.

Adult midges are probably easy to kill with insecticides. A problem is that the flies lay their eggs shortly after they emerge, possibly on the same day, and the eggs are deposited in crevices of the flower buds or leaves, where they may be sheltered from insecticides. Insecticides might also be found that would kill the larvae that are believed to spend the summer and fall in the top 2 centimeters of soil beneath the plants. Research is needed on the optimum timing of insecticide sprays to kill the adults. One strategy would be to protect the developing flower buds, targeting the sprays for each cultivar to match the 2-to-3-week interval when the flower buds of that cultivar begin expanding and are most susceptible to midges. This would require monitoring the stage of flower bud development for each cultivar and determining each day during the vulnerable period whether fly populations are high enough to require spraying. Limited tests in commercial fields in Florida suggest that this strategy can be effective. A second strategy would be to apply an insecticide to kill the flies that infest vegetative meristems after bees have been removed from the field. Observations indicate that huge population increases can occur at this time. The goal would be to reduce the number of over-summering larvae that would emerge to threaten flower buds the following winter. Growers in southeast Georgia who spray after flowering to control cranberry fruitworm have reported reductions in gall midge damage on vegetative meristems.

Breeding resistant or tolerant rabbiteye cultivars is undoubtedly possible, in view of the fact that 'Powderblue' and 'Brightwell' and most southern highbush cultivars appear to be highly resistant. Problems with this approach are that cultivar development takes many years, most rabbiteye cultivars are moderately to highly susceptible, and many other plant characteristics also demand the attention of the breeder. Another concern is that the advent of new biotypes of the insect could cause resistance to break down. This has occurred repeatedly with Hessian fly, a species of gall midge that attacks wheat (Buntin, et al., 1990). In a long-lived perennial crop such as blueberries, loss of resistance could be disastrous. On the other hand, the apparent stability and near universality of flower bud resistance to gall midge in highbush blueberry suggests that breeding stable resistance in rabbiteye may be possible.

In wheat, 20 different genes, mostly non-allelic, have been found to confer resistance to Hessian fly (Raupp et al., 1993). All genes but one are partially to completely dominant in expression. The mechanism of resistance for all genes is larval antibiosis, with first instars dying after feeding on resistant plants (Raupp et al., 1993). Keep (1985) reported evidence for a dominant gene for resistance to black currant (*Ribes nigrum* L.) leaf-curling-midge (*Dasineura tetensi* Rubs) in *Ribes dikuscha*. Two Scandinavian black currant cultivars were also highly resistant to the midge.

GROWER BENEFITS

Flower bud damage from the blueberry gall midge makes production of rabbiteye blueberries commercially infeasible in the Florida peninsula south of Gainesville. Now that the problem is recognized, it should be possible to devise methods to reduce the damage, either through controlling the pest or through development of resistant cultivars. This might make it possible to grow low-chill rabbiteye blueberries in south-central Florida to satisfy the potential pick-your-own market and to provide a blueberry that can be mechanically harvested for the fresh market from late April through May.

LITERATURE CITED

Barnes, H.F. 1948. Gall Midges of Economic Importance. III. Gall Midges of Fruit. Crosby, Lockwood, and Son. London.

Buntin, G.D., P.L. Bruckner, J.W. Johnson, and J.E. Foster. 1990. Effectiveness of selected genes for Hessian fly resistance in wheat. J. Agric. Entomol. 7:283-291.

Gagné, R.J. 1989. The Plant-Feeding Gall Midges of North America. Cornell University Press, Ithaca, New York.

Keep, E. 1985. The black currant leaf curling midge, *Dasyneura tetensi* Rubs.; its host range, and the inheritance of resistance. Euphytica 34:801-809.

Lyrene, P.M. and J.A. Payne. 1992. Blueberry gall midge: a pest on rabbiteye blueberry in Florida. Proc. Fla. State Hort. Soc. 105:297-300.

Mahr, D.L. 1991. Cranberry tipworm: preliminary results of 1990 sanding studies. 1991 Proc. Wisconsin Cranberry School. pp. 45-48.

Mahr, D.L. and R. Kachadoorian. 1990. Cranberry tipworm. 1990 Proc. Wisconsin Cranberry School. pp. 17-22.

Mainland, C. M., J.T. Ambrose, and L.E. Garcia. 1979. Fruit set and development of rabbiteye blueberries in response to pollinator, cultivar, or gibberellic acid. In: J. N. Moore (ed.). Proc. 4th North American Blueberry Research Workers Conf. Univ. of Arkansas, Fayetteville. pp. 72-73.

Payne, J.A., A.A. Amis, R.J. Beshear, R.J. Gagné, D.L. Horton, and P.M. Lyrene. 1993. New rabbiteye blueberry insects: maggots, midges, thrips, and root weevils. Proc. 6th Biennial Southeast Blueberry Conference, Tifton, Georgia.

Raupp, W.J., A. Amri, J.H. Hatchett, B.S. Gill, D.L. Wilson, and T.S. Cox. 1993. Chromosomal location of Hessian fly-resistance genes H22, H23, and H24 derived from *Triticum tauschii* in the D genome of wheat. J. Hered. 84:142-145.

Scammell, H.B. 1917. Cranberry insect problems and suggestions for solving them. Farmers' Bulletin 860. U.S. Dept. Agriculture, Washington.

Smith, J.B. 1903. Insects injurious in cranberry culture. Farmers' Bulletin 178. U.S. Dept. Agriculture, Washington.

Spiers, J.M. Effect of stage of bud development on cold injury in rabbiteye blueberry. J. Amer. Soc. Hort. Sci. 103:452-455.

Vanderwegen J. and G. Krewer. 1991. Gibberellic acid has potential for use in rabbiteye blueberries. In: Proc. 5th Biennial Southeast Blueberry Conference, Tifton, Georgia. pp. 23-28.

Control of Bunchberry
in Wild Blueberry Fields

David E. Yarborough
Timothy M. Hess

SUMMARY. Bunchberry has been increasing in Maine and Canadian wild blueberry fields with the use of the herbicide hexazinone.
A nonbearing year Spring application of tribenuron methyl suppressed bunchberry without yield loss to wild blueberry. A combined
Spring and Fall application resulted in unacceptable injury to blueberry growth and yield. The use of a surfactant did not improve
bunchberry control. *[Article copies available from The Haworth Document Delivery Service: 1-800-342-9678.]*

KEYWORDS: *Vaccinium angustifolium*; *Cornus canadensis*; Tribenuron methyl; Express

INTRODUCTION

Bunchberry (*Cornus canadensis* L.) has invaded spaces in and
among commercial wild blueberry (*Vaccinium angustifolium* Ait.)
clones. Bunchberry invades barren areas left open by applications

David E. Yarborough is Assistant Professor of Horticulture; Timothy M. Hess
is Research Associate, Department of Applied Ecology and Environmental
Sciences, 5722 Deering Hall, University of Maine, Orono, ME 04469-5722.
Maine Agricultural and Forest Experiment Station Contribution No. 1899.

[Haworth co-indexing entry note]: "Control of Bunchberry in Wild Blueberry Fields." Yarborough,
David E., and Timothy M. Hess. Co-published simultaneously in *Journal of Small Fruit & Viticulture*
(Food Products Press, an imprint of The Haworth Press, Inc.) Vol. 3, No. 2/3, 1995, pp. 125-132; and:
Blueberries: A Century of Research (ed: Robert E. Gough, and Ronald F. Korcak) Food Products Press,
an imprint of The Haworth Press, Inc., 1995, pp. 125-132. Single or multiple copies of this article are
available from The Haworth Document Delivery Service [1-800-342-9678, 9:00 a.m. - 5:00 p.m. (EST)].

of Velpar® (hexazinone) more rapidly than blueberries (Yarbo-rough and Bhowmik, 1993) and produces an undesirable fruit. McCully et al. (1991) found bunchberry to be the most prevalent weed after the use of Velpar in Nova Scotia blueberry fields. Bunchberry has a similar growth pattern to blueberry making selective control difficult (Yarborough and Bhowmik, 1989). While Roundup® (glyphosate) may be used to suppress pure stands of bunchberry, there is no selective measure once bunchberry begins growing within the blueberry clones. Earlier trials (Yarborough, 1992; Yarborough and Bhowmik, 1989) indicated the sulfonyl urea class of herbicides could be effective in suppressing bunchberry.

The objective of this experiment was to test the effectiveness of Express® (tribenuron methyl) with or without surfactant at different rates and dates, to suppress bunchberry and to assess its effect on the growth and yield of the wild blueberry.

MATERIALS AND METHODS

Two split block experiments were set out at Blueberry Hill Farm, Jonesboro, ME on a Colton sandy loam soil in May 1991 and 1992. In 1991, the experiment had four rates of Express, 0, 0.14, 0.28 and 0.57 oz ai/a with or without surfactant, applied on June 20 or July 14 on five replications for a total of 80 plots. In 1992, the experiment had the same rates without surfactant but applied on May 29 or June 17 or on May 29 and September 14 on six replications for a total of 72 plots. Plot size was 3 by 16 ft with 3 by 16 ft alleyways and two, 1 ft^2 count plots per plot. Pre-treatment bunchberry counts were taken in May of the treatment year and phytotoxicity was evaluated using a subjective rating scale of 0 to 100 with 0 having no injury and 100 being dead. Number of remaining bunchberry were counted in June of the crop year and plots were harvested in August. Data were analyzed by analysis of variance using the SAS GLM procedure (SAS, 1985).

RESULTS AND DISCUSSION

In both years the effectiveness of Express in reducing bunchberry numbers increased with rate of application (Figure 1). The June application was effective in reducing bunchberry in both years but the combined spring-fall gave the greatest suppression. Blueberry phytotoxicity was observed with both the spring and fall applications. The June 20 application resulted in a stunted, reddened appearance but the combined Spring and Fall application resulted in excessive necrosis to the blueberry and an absence of bloom (Figure 2). No differences were found in effectiveness or phytotoxicity was obtained with the surfactant. In 1991 increasing the rate of Express stimulated blueberry flower bud production but no significant effect was observed with the 1992 application (Figure 3). Blueberry yield was reduced with increasing rates of Express in 1991 but not significantly reduced by rate of Express for the May 29 or June 17 applications in 1992 (Figure 4). However, the blueberry yield for the 1991 July 14 application was significantly less than the June 20 application. The data for the combined spring and fall 1992 applications showed a significant decline in blueberry yield with higher rates of Express (Figure 4).

CONCLUSION

Express applied in early Spring at 0.14 to 0.57 oz ai/a effectively reduced bunchberry number without significantly injuring blueberry or reducing blueberry yield. McCully (1994) indicated blueberry plant stunting could occur from Express rates of 0.57 oz ai/a but the plants recovered and there was no effect on yield. This study confirms his results. Since bunchberry is not a competitor with blueberry (Yarborough and Bhowmik, 1993) an immediate increase in yield was not expected. However, if bunchberries are not in the interclonal spaces, blueberry plants will fill in and yield will increase over time. Later spring or fall applications resulted in significant blueberry injury and reduction in yield.

GROWER BENEFIT

Early Spring applications of Express will selectively control bunchberry in wild blueberry stands. Control of bunchberry will allow blueberry stands to fill in and increase the long-term yields in wild blueberry fields.

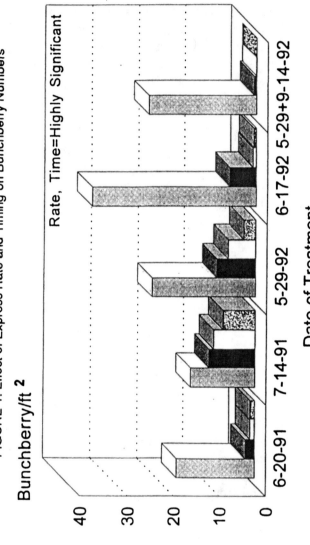

FIGURE 1. Effect of Express Rate and Timing on Bunchberry Numbers

128

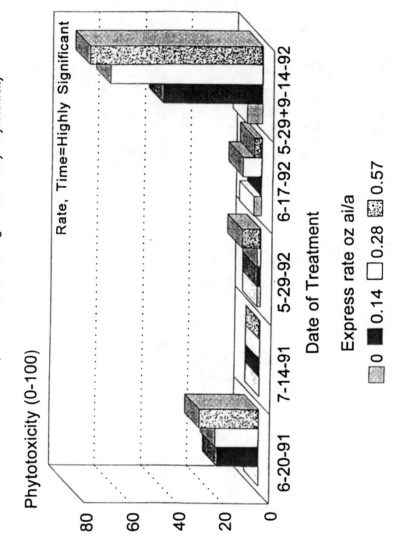

FIGURE 2. Effect of Express Rate and Timing on Blueberry Phytotoxicity

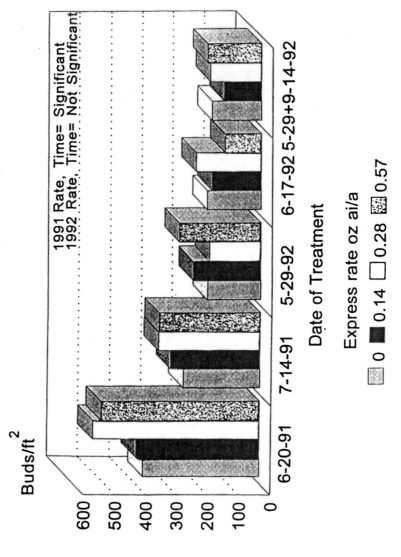

FIGURE 3. Effect of Express Rate and Timing on Blueberry Flower Buds

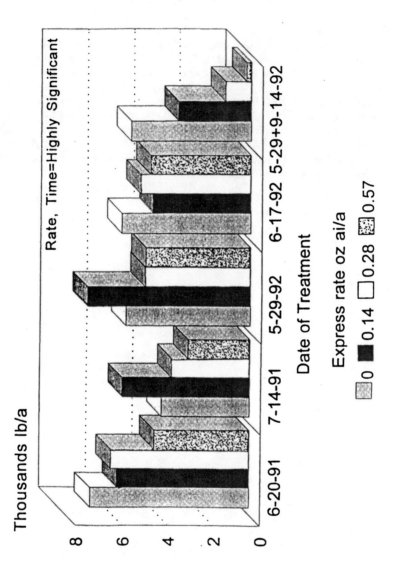

FIGURE 4. Effect of Express Rate and Timing on Blueberry Yield

LITERATURE CITED

McCully, K.V., M.G. Sampson, and A.K. Watson. 1991. Weed survey of Nova Scotia blueberry (*Vaccinium angustifolium*) fields. Weed Sci. 39:180-185.

McCully, K.V. 1994. Spring registration of tribenuron methyl (Express). Proceedings New Brunswick Horticultural Conference Technology and Marketing. pg. 10.

SAS Institute 1985. SAS User's Guide: Statistics. Version 5. Cary, N.C.

Yarborough, D. E. 1992. Evaluation of sulfonyl urea herbicides for bunchberry suppression in lowbush blueberry fields. Proc. Northeast. Weed Sci. Soc. 46:21.

Yarborough, D.E. and P.C. Bhowmik. 1989. Evaluation of sulfonyl urea and imidazoline compounds for bunchberry control in lowbush blueberry fields. Proc. Northeast. Weed Sci. Soc. 43:142-145.

Yarborough, D.E. and P.C. Bhowmik. 1989. Effect of hexazinone on weed populations and on lowbush blueberries in Maine. Acta Hort. 241:344-349.

Yarborough, D.E. and P.C. Bhowmik 1993. Lowbush blueberry-bunchberry competition. J. Amer. Soc. Hort. 118:54-62.

RELATED POSTERS

Influence of Mulching Systems on Yield and Quality of Southern Highbush Blueberries

J. B. Magee
J. M. Spiers

SUMMARY. The effects of four mulching systems (pine bark, black woven fabric, black plastic, white-over-black plastic) on plant growth, yield and fruit quality of five southern highbush blueberry clones were studied. Plant growth and fruit yields of plants grown on pine bark or white-over-black plastic were not different and were significantly higher than growth and yield of plants grown on the black mulches. The mulching systems did not significantly affect berry weight, soluble solids, titratable acidity, soluble solids/acid ratio, anthocyanins, skin firmness or subjective quality after three weeks of refrigerated storage. There were significant effects of clone on each of these parameters. *[Article copies available from The Haworth Document Delivery Service: 1-800-342-9678.]*

J. B. Magee and J. M. Spiers, Research Horticulturists, USDA-ARS Small fruit Research Station, P.O. Box 287, Poplarville, MS 39470.

[Haworth co-indexing entry note]: "Influence of Mulching Systems on Yield and Quality of Southern Highbush Blueberries." Magee, J. B., and J. M. Spiers. Co-published simultaneously in *Journal of Small Fruit & Viticulture* (Food Products Press, an imprint of The Haworth Press, Inc.) Vol. 3, No. 2/3, 1995, pp. 133-141; and: *Blueberries: A Century of Research* (ed: Robert E. Gough, and Ronald F. Korcak) Food Products Press, an imprint of The Haworth Press, Inc., 1995, pp. 133-141. Single or multiple copies of this article are available from The Haworth Document Delivery Service [1-800-342-9678, 9:00 a.m. - 5:00 p.m. (EST)].

133

KEYWORDS: Southern highbush blueberries; *Vaccinium corymbosum* hybrids; Mulching systems; Blueberry quality

INTRODUCTION

The favorable economic potential for early market blueberries has caused growers of rabbiteye blueberries (*Vaccinium ashei* Reade) in the Gulf Coast states to consider planting southern highbush blueberries (hybrids, mainly *V. corymbosum* L.). Southern highbush (SH) blueberries have low chilling requirement and short fruit development periods. In this area SH blueberries ripen two-four weeks earlier than rabbiteye cultivars which allows growers to take advantage of potentially higher-profit market windows. In 1992 at Poplarville, Mississippi, the SH cultivars 'Gulfcoast' and 'Cooper' were 50% ripe, May 16 and 19 respectively, 10 and 13 days earlier than 'Climax' and a month earlier than 'Tifblue,' two rabbiteye cultivars widely grown in this area (Gupton, 1993).

Mulched SH plants are larger and higher yielding than unmulched plants under southern growing conditions (Clark and Moore, 1991; Spiers, 1992). Pine bark, wood chips or pine sawdust are often used as mulches in the Gulf States area because of their availability. Commercial mulching products or ground covers are used with other small fruits and vegetables, but information on their use with SH blueberries is lacking. The objective of this study was to determine the influence of four mulching systems on plant growth, berry yield and fruit quality parameters of six SH cultivars or selections (clones).

MATERIALS AND METHODS

Two-year-old container-grown plants were planted in April 1990 at the Small Fruit Research Station, Poplarville, Mississippi, in a well-drained fine sandy loam soil. The SH clones 'Gulfcoast,' 'Cooper,' MS132, MS161, and MS179 were established in a randomized complete block design with four treatments (mulches), five plants/treatment and four blocks (rows). The beds were tilled and treated with methyl bromide. Twelve L of peatmoss per plant were incorporated into the middles of the beds. Plant spacing was 0.9 m (3 ft) in rows 3 m (10 ft) apart. In April 1991 four mulching systems (pine bark and three commercial ground covers) were installed around the plants. The commercial ground covers were black woven fabric, black plastic and white-over-black (W/B) plastic. Width of each mulch was 1.2 m (4 ft), and the pine bark depth was approximately 15.2 cm (6 in). Standard cultural practices for blueberries in this area were followed, and trickle irrigation with a single drip emitter per plant was used as needed.

Plant size (volume) was estimated by multiplying width × spread × height of each plant (Clark and Moore, 1991). Soil temperature was measured 10 cm (4 in) below the mulch-soil interface.

Fruit was harvested beginning at approximately 30% ripeness of each plant and at four- or five-day intervals thereafter, weather permitting. Harvested fruit was weighed and sorted to remove soft, damaged or unripe berries. The weight of 30 berries was used to calculate an average berry weight. The force required to penetrate the epidermis was measured with a gram gauge penetrometer (Perkins-Veazie et al., 1993) fitted with a blunt 0.316 mm diameter probe. Three penetrometer values were taken on each of 30 berries (stem shoulder, blossom shoulder and equatorial), and the 90 values were averaged and reported (Newtons/m^2 × 10^4) as an estimation of skin firmness. For estimating storage quality, samples of berries were held in 473 ml (1 pint) clear plastic "clam shell" containers at 4°C (39°F) for 20 days. Weight lost during storage was calculated. Storage quality was rated by assigning subjective values (1-5 with 1 = good and 5 = poor) for shrivel, softness and decay of 10-berry subsamples; these values were averaged and reported as a storage rating. Containers of berries with a storage rating of 3 or above would be unsalable.

Fruit pH, titratable acidity (TA), soluble solids concentration (SSC) and anthocyanin (ACY) concentration were determined using fruit frozen at harvest; samples from each harvest were pooled for analyses. Forty g of fruit was homogenized with a Waring blender, and the pH of the homogenate was measured. TA was determined by diluting 5 g of homogenate to 100 ml with deionized water and titrating to a pH of 8.2 with 0.1 N NaOH; TA was expressed as percent citric acid. SSC was measured with a hand-held refractometer using liquid from the homogenate filtered through four layers of grade 50 cheesecloth. To estimate ACY, 5 g of homogenate was extracted twice with 95% reagent ethanol:0.1 N HCl (85:15). The supernatants were combined and diluted to 50 ml with extraction solution. Absorbance at 533 nm was measured with a Perkin-Elmer Lambda 3B spectrophotometer. ACY content was expressed as absorbance units per g of fresh weight (AU/g).

RESULTS AND DISCUSSION

Total yield and plant volume were the only dependent variables affected by the mulching treatments. In 1994 fruit yield and volume of plants grown on W/B plastic and pine bark were significantly higher than those of plants on black plastic or black fabric (Table 1). Yield data were taken on the small first crop in 1992; plants on pine bark had a slightly higher yield than plants on W/B plastic, and both yields were higher than those of plants on the black mulches. Data were not taken in 1993 because a severe freeze in mid-March killed most of the crop.

Soil moisture was not measured in this study, but differences in water available to the plants probably were not the major reason for the differences in plant growth and berry yields. Plants grown on

TABLE 1. Effect of mulching treatments on yields and plant volumes.

Treatment	Yield, g		Plant volume, m^3
	1992	1994	1994
Bark	375	3222 a^z	0.18 a
W/B plastic	350	3378 a	0.17 a
Black fabric	268	2224 b	0.13 b
Black plastic	268	2123 b	0.12 b

^{z}Means with the same letter are not different at $P < 0.05$, Duncan's multiple range test.

the two plastic mulches should have responded similarly in their interactions with irrigation water, soil moisture conservation or gas exchange, yet the growth and berry yields of their respective plants were significantly different (Table 1). Pine bark and the fabric mulch are permeable to gas and water exchange, but the growth and yields of the plants in pine bark was significantly higher than those on black fabric.

These results differ from those reported by Creech (1990) for two rabbiteye cultivars. In a three-year establishment study he evaluated the effects of three above-ground treatments (continuous pine bark mulch, black polyfabric and none) and four in-ground amendment treatments with 'Brightwell' and 'Climax.' After three years in the field, plants mulched with polyfabric were significantly larger and had larger root systems than plants mulched with pine bark or not mulched. He noted that all root systems were shallow with most roots in the top 20.3 cm (8 in) of soil and that there was intense proliferation of fibrous roots immediately under the mulches whether grown under polyfabric or bark. He suggested likely explanations for the better growth of the plants on polyfabric might include less nitrogen competition vs. bark, a broader wetting pattern and a warmer springtime temperature. He attributed the good root growth under the polyfabric to factors other than organic matter.

SH roots may be more sensitive to high soil temperatures than rabbiteye roots (Spiers, 1992). A higher percentage of SH roots were found in the top 15.2 cm (6 in) of soil and the lateral spread of SH roots was smaller when compared to rabbiteye roots (Spiers, 1986 and Makus et al., 1994). The shallow depth and small spread of SH roots could leave them more vulnerable to adverse effects of higher soil temperatures under the black mulches which could result in reduced plant growth and fruit yields. Average 3:00 pm soil temperatures 10.2 cm (4 in) below the mulch-soil interfaces (Table 2) illustrate the effects of an insulating mulch (bark) and a reflective/opaque mulch (W/B plastic) when compared with the more heat absorbing black mulches. In the first year of his study, Creech (1987) reported that a slight film of sand drifted over the black polyfabric and lowered its surface temperature; this may have afforded some protection to the roots. The root systems of our plants will be examined at the end of the study.

TABLE 2. Average soil temperatures four inches under mulches at 3:00 p.m., 1992.

Treatment	Apr 13-May 15	Jul 15-Aug 15
Bark	21.5	31.2
W/B plastic	22.5	38.4
Black fabric	24.9	40.8
Black plastic	25.8	44.2
Soil, no mulch	27.6	37.0
Air	21.4	33.8

There were no significant treatment effects on fruit quality parameters, but there were differences due to clone (Table 3). Averaged across treatments, the yield of MS132 was much higher than that of the other clones, almost double that of the next highest. The plant volume of MS132 also was larger than those of the other clones.

MS161 had the smallest plant volume, highest berry weight, lowest ACY concentration, highest pH and lowest TA (Table 3). Because of its low TA, the SSC/TA ration of MS161 was much higher than that of the other clones. A higher SSC/TA ratio has been associated with higher incidences of storage decay in both highbush and rabbiteye blueberries (Ballinger and Kushman, 1970; Miller et al., 1988), but in this study MS161 was the only clone whose berries were rated salable after three weeks' storage. There were no differences among clones in storage ratings at one and two weeks.

CONCLUSION

Under the conditions of this study, plant growth and berry yield of five SH clones grown on pine bark or white-over-black mulches were not different but both were significantly higher than the

TABLE 3. Effect of clone on the values of fruit quality parameters.

Clone	Yield (g)	Plant volume (m³)	Berry weight (g)	pH	SSC (%)	TA (% as citric)	SSC/TA	ACY (AU/g fw)	Skin firmness (N/m² × 10⁴)	Storage rating 3 weeks
'Cooper'	2016 bᶻ	0.16 b	1.28 b	2.71 bc	8.6 b	1.05 a	8.4 b	27.0 a	163 a	3.3 a
'Gulfcoast'	1795 b	0.14 b	1.16 b	2.77 b	8.4 b	0.94 ab	9.2 b	26.8 ab	164 a	3.4 a
MS132	5175 a	0.20 a	1.73 a	2.70 bc	10.1 a	1.01 ab	10.4 b	23.6 bc	151 b	3.2 a
MS161	1815 b	0.08 c	1.90 a	3.10 a	8.5 b	0.31 c	34.8 a	14.5 d	146 b	2.9 b
MS179	2683 b	0.14 b	1.28 b	2.65 c	9.9 a	0.88 b	12.2 b	23.3 c	170 a	3.3 a

ᶻMeans with the same letter are not different at P < 0.05, Duncan's multiple range test.

139

growth and yield of plants grown on black woven fabric or black plastic. We attribute most of these differences to the adverse effects of higher root zone soil temperatures under the two black mulches. The mulch treatments did not significantly affect berry weight, soluble solids concentration, titratable acidity, soluble solids/acid ratio, anthocyanin concentration, skin firmness or subjective quality after three weeks' refrigerated storage.

GROWER BENEFITS

Growers in the Gulf States who plant southern highbush blueberries should mulch for increased plant growth and berry yield. Where available and economically feasible, traditional organic mulches such as pine bark can be used. Light-reflecting but opaque ground covers such as white-over-black plastic may be useful as mulch substitutes, but more information on their effectiveness is needed. Mulching materials that increase the root-zone soil temperature, such as black plastic, should not be used.

LITERATURE CITED

Ballinger, W. E. and L. J. Kushman. 1970. Relationships of stage of ripeness to composition and keeping quality of highbush blueberries. J. Amer. Soc. Hort. Sci. 95:239-242.

Clark, J. R. and J. N. Moore. 1991. Southern highbush blueberry response to mulch. HortTechnology 1:52-54.

Creech, D. 1987. The influence of three above-ground mulch treatments and four in-ground amendment treatments on growth of 'Climax' and 'Brightwell' rabbiteye blueberries. Proc. Texas Blueberry Growers Assn. Ann. Conv., Nacogdoches, TX, pp. 66-72.

Creech, D. 1990. Final report on the polyfabric, bark and zero aboveground mulch study. Proc. Texas Blueberry Growers Assn. Ann. Conv., Nacogdoches, TX, pp. 40-48.

Gupton, C. L. 1993. Update on blueberry variety research in Mississippi. Proc. 6th Biennial Southeast Blueberry Conf., Tifton, GA, pp. 13-17.

Makus, D. J., J. M. Spiers, K. D. Patten and E. W. Neuendorff. 1994. Growth responses of southern highbush and rabbiteye blueberry cultivars at three southern locations. Presented at 7th N. Amer. Blueberry Research-Extension Workers Conf., Beltsville, MD.

Miller, W. R., R. E. McDonald and T. E. Crocker. 1988. Fruit quality of rabbiteye blueberries as influenced by weekly harvest, cultivars, and storage duration. HortScience 23:182-184.

Perkins-Veazie, P. M., J. H. Collins and J. R. Clark. 1993. Fruit characteristics of some erect blackberry cultivars. HortScience 28:853.

Spiers, J. M. 1992. Establishment of 'Gulfcoast' southern highbush blueberry. HortScience 27:148. (Abstract).

Spiers, J. M. 1983. Irrigation and peatmoss for the establishment of rabbiteye blueberries. HortScience 18:936-937.

Spiers, J. M. 1986. Root distribution of 'Tifblue' rabbiteye blueberry as influenced by irrigation, incorporated peatmoss and mulch. J. Amer. Soc. Hort. Sci. 111:877-880.

Operations and Cost
of Highbush Blueberry Farming in Poland

T. Marzec-Wołczyńska
D. Karwowska

SUMMARY. The economic conditions in Poland for growers of agricultural products have become less favorable since the frontiers have been opened to international competition. One of the lines of fruit growing which still remains profitable is production of blueberry fruit and young plants. It is so because the costs of production are relatively low, especially the costs of hand labor. Another favorable condition is the growing demand. A limiting circumstance is the difficulty in obtaining credit and the extremely high interest. The present paper is based on data from a well-managed blueberry farm. It brings a description of the farm and unique (so far unpublished) information on the level and structure of production costs and the size of income. *[Article copies available from The Haworth Document Delivery Service: 1-800-342-9678.]*

INTRODUCTION

Poland was the second country (after Holland) in Europe into which the American highbush blueberry was introduced in the

T. Marzec-Wołczyńska, Department of Horticultural Economics, Warsaw Agricultural University-SGGW, Nowoursynowska 166, 02-766 Warszawa, Poland.

D. Karwowska, Blueberry Farm "DAAR," Piskórka 10, 05-540 Zalesie Górne, Poland.

[Haworth co-indexing entry note]: "Operations and Cost of Highbush Blueberry Farming in Poland." Marzec-Wołczyńska, T., and D. Karwowska. Co-published simultaneously in *Journal of Small Fruit & Viticulture* (Food Products Press, an imprint of The Haworth Press, Inc.) Vol. 3, No. 2/3, 1995, pp. 143-149; and: *Blueberries: A Century of Research* (ed: Robert E. Gough, and Ronald F. Korcak) Food Products Press, an imprint of The Haworth Press, Inc., 1995, pp. 143-149. Single or multiple copies of this article are available from The Haworth Document Delivery Service [1-800-342-9678, 9:00 a.m. - 5:00 p.m. (EST)].

143

1920s, but its cultivation did not expand. After World War II, new plants were imported, but it was only as late as 1970 that large scale research in the Department of Horticulture of the Warsaw Agricultural University and widespread growing were undertaken. From the imported material the first stage-managed plantings were started, some of them still in operation to this day. Private enterprises and companies producing young material started a few years later.

It is estimated that at present there exist in Poland about several score of plantings with a total of about 200 hectares (494 A). Three of them exceed 20 ha (48 A), and the remaining ones are about 3 ha (7 A) each. The production of nurseries is estimated to be around 200,000 plants annually, and fruit production is about 350 to 400 tons. The demand for the fruit in Poland is high since the blueberry harvest starts at a time when there is a shortage of native fruits. The prices on the home market are more attractive than those for export; however, since the distribution channels have so far not been well organized, most of the fruit production is exported.

The Polish producers are dispersed; they have no professional or marketing organization. Thus, detailed information is lacking on the size of production, sales, costs and profits. It should also be stressed that in Poland farmers and gardeners do not pay income taxes and bookkeeping is not obligatory for them. The information in the present paper on the organization and costs of production is derived from one blueberry farm collaborating with the Department of Pomology (Warsaw Agricultural University). These data are unique; they have not been published anywhere and were elaborated for presentation at the Symposium.

MATERIALS AND METHODS

This paper is a description of a Polish farm growing blueberries exclusively. It is a private (family) enterprise, and is the only source of income of a family of five. The owners of the farm, a couple, are both horticulture engineers and manage the planting in consultation with the Department of Pomology. They keep records of income and expenses for their own use. These records for the year 1993 served as basis for the present elaboration.

In presentating costs, the system from the "Highbush Blueberry Production Guide" (Pritts and Hancock, 1992) was adopted, in order to facilitate comparison with American data. The variable costs are subjected to detailed analysis. The true value of fixed costs is difficult to calculate. The farm has been existing for 14 years and is continuously developing. Over this period the prices of land, machines and buildings have drastically changed in Poland; therefore, there was an attempt to jointly estimate the sum of fixed costs in a farm of this type according to current prices.

RESULTS AND DISCUSSION

The farm lies 30 km (18.6 mi) south of the city of Warsaw. Its surface area is 10 ha (24.7 A). This includes 8 ha (19.8 A) of planting for fruit and 0.3 ha (0.74 A) of mother stock. The farm has a residential house and a farm building. Equipment includes: two garden tractors, 3 trailers, spraying machine, knapsack sprayer, two mowing machines, pallet elevator, 3 foil tents of 180 m^2 each, cold store 0°C (32°F) of 5 ton volume, installation for irrigation (micro-spray), automatic installation for irrigation and warming of soil tents and installation for irrigating the nursery.

The farm is managed by the owner couple; moreover, they employ three persons permanently, and seasonally for pruning, ten for technical practices during 8 weeks, and thirty during harvest for 6 weeks.

The main activity is the growing of young plants in the nursery. Semi-ripewood cuttings are rooted in single-span polyethylene-clad propagation greenhouses, in bottom-heated substrate. After the winter the young plants are transplanted into 1.5 L containers and in the fall they reach a commercial size. They are mainly sold in Poland: fifty per cent to small private gardens by gardening shops. The rest is purchased for new plantings.

The fruits are becoming a more and more important item for profit. The surface of the plantation is 8 ha (19.8 A), the age of the bushes is from 3 to 10 years, the spacing is 3 × 1 m (10 × 3 ft), the mid-spaces are mulched with sawdust and alleyways are kept in sod. The planting and the nursery are equipped with installations for fertigation. 'Bluecrop' constitutes 80 per cent of the plantings; the remaining cultivars used are: Bluetta, Bluejay, Berkeley, Jersey and Darrow. Other plantings have a similar cultivar structure. There are two reasons for this: first of all 'Bluecrop' is above competition as a cultivar, secondly others ('Berkeley' excepted) have so far not confirmed their high qualities in our climate. In the foregoing year, testing of new cultivars brought from the USA, such as 'Sunrise,' 'Spartan,' 'Patriot,' 'Nelson' and 'Bluegold,' was started. This will prevent periodic peaks of availability and fall of prices. It is particularly important in Poland, where all the American blueberries are consumed fresh.

With increasing harvests there arises the need of an infrastructure which would serve the sales. The existing cool storage space was already too small in the preceding year. In the present season the purchase of a new storage and a sorting line is planned. All this is not simple, since installations and machines are not produced in Poland for management of highbush blueberries. This equipment has to be purchased in Western Europe or even in America.

The farm has been in operation since 1980. In 1993 the joint profit from sale of young plants was about US$ 120,000 (60,000 plants–US$ 90,000; 15 tons fruits–US$ 30,000).

The demand for plants continues to be high. The cost of one young bush is US$ 1.50. The fruit is mostly exported to Great Britain (80%). The export prices are close to the local ones. In 1993 the cost of 1 kg of the fruit was around US$ 2.00 in Poland, and for export, US$ 1.70.

The variable costs of fruit production is shown in Table 1, and for planting material in Table 2. The costs of fruit production as compared with the data in the Hudson Valley Study (Pritts and Hancock, 1992) are lower by almost one-half. Analysis of the structure of investments indicates that the differences concern mainly the price of labor, especially qualified labor (repair of machines, tractors, administration). The use of hand labor in Poland is much higher. This is connected with labor organization, for most of the work is done by hand. In addition to variable costs, each hectare of plantings is burdened with an annual amount of about US$ 2,500 of fixed costs.

This situation concerns the period when unemployment in Poland is high, and at the same time the remuneration possible to

TABLE 1. Variable production costs of fruit section per acre (year 5).

COSTS	UNIT	PRICE or COST/UNIT	QUANTITY	VALUE or COST
PREHARVEST				
Labor hand pruning	hour	1.50	40	60
Ammonium sulfate	pound	0.10	110	11
Labor apply A.S.	hour	1.50	1	1.50
Sawdust	m^3	2.50	60	150
Hand applied mulch	hour	1.50	6	9
Irrig. equip. startup	hour	1.50	1	1.50
Euparen	pound	10.00	1.3	13
Irrig. operating	acre	20.00	1	20
Management	acre	50.00	1	50
Tractors repair	acre	10.00	1	10
Tractors fuel/lube	acre	5.50	1	5.50
Machinery repairs	acre	3.00	1	3.00
Labor (tractor/machinery)	hour	1.50	11	16.50
SUBTOTAL, PREHARVEST				$ 351.00
HARVEST				
Labor harvest prep.	hour	1.50	3	4.50
Hand harvest	pound	0.08	6500	520
Labor clean/up	hour	1.50	80	120
Labor (tractor/machinery)	hour	1.50	4	6
SUBTOTAL, HARVEST				$ 650.50
POSTHARVEST				
Hand hoeing	hour	1.50	5	7.50
NPKMg	pound	0.09	80	7.20
Labor NPKMg	hour	1.50	1	1.50
Roundup	liter	15.00	2	30.00
Tractors Repair	acre	7.00	1	7.00
Tractors Fuel/Lube	acre	3.50	1	3.50
Labor (Tractor/Machinery)	hour	1.50	6	9.00
SUBTOTAL POSTHARVEST				$ 65.70
TOTAL VARIABLE COST				$ 1067.20

TABLE 2. Variable costs of nursery section per 100 thousands rootings in USD (plant price US$ 1.50).

COSTS	UNIT	PRICE or COST/ UNIT	QUAN- TITY	VALUE or COST	VALUE in YOUNG PLANTS
1. Peat	m³	9.60	40	385.50	257
2. Sand	m³	5.25	20	105.00	70
3. Semi-ripewood cuttings	m³	0.04	140000	6000.00	4000
4. Peat	m³	9.6	50	480.00	320
5. Sand	piece	5.25	40	210.00	140
6. Ground bark	m³	3.40	70	240.00	160
7. Pots	piece	0.01	100000	1000.00	660
8. Plantacote	pound	1.05	1000	1050.00	700
9. Ammonium Sulfate	pound	0.10	500	50.00	33
10. Euparen	pound	10.00	5	50.00	33
11. Labels and leaflets	piece	0.15	50000	7500.00	5000
12. Labor					
– substrate and bed preparation	hour	1.50	48	72.00	48
– planting of cuttings	hour	1.50	900	1350.00	900
13. Labor: planting					
– substrate and bed preparation	hour	1.50	120	180.00	120
– transporting	hour	1.50	1800	2700.00	1800
– care of polyethylene houses	hour.	1.50	900	1350.00	900
and nursery	hour	1.50	2160	3240.00	2160
– preparation of young plants	hour	1.50	720	1080.00	720
TOTAL				$ 27042.50	18021

obtain in agriculture is low. It should be mentioned that seasonal laborers in agriculture are not insured in Poland. Thus, this part of costs is relatively low. We are, however, at the start of introduction of a market economy and changes in the system of wages and taxes. As a result of this, within the next few years the structure of production costs may change dramatically.

At present, it is estimated that blueberry production in Poland is one of the most profitable branches of fruit growing. At the same time, however, it is a capital-consuming production. The high costs of establishing a planting, under conditions of poor availability and high expenses of credit, inhibit new investments.

CONCLUSION

A farm specializing in the growing of highbush blueberry in Poland on an area of 10 ha (24.7 A) can be the source of quite a good income for a family. The structure of costs indicates that a high proportion of them consists of expenses for hand labor which are much higher than, for instance, in the USA. The price of labor is at present relatively low, but an increase of wages and burdens connected with this is to be expected.

GROWER BENEFITS

Economic analysis of the production of highbush blueberry fruits and planting material indicates that it may be a good alternative to orchard operations which have ceased to be profitable, such as black currants or sour cherry production. The demand for both fruits and young blueberry plants is increasing. The economics of new plantations can be profitably influenced by their suitable localization. Large capital investments are, however, indispensable. The approach for obtaining bank credits may be facilitated by an analysis of the costs according to the examples shown in this publication.

LITERATURE CITED

Pritts M. P. and J. F. Hancock. 1992. Highbush Blueberry Production Guide, Cooperative Extension, Ithaca, NY. NRAES-55 pp. 156-168.

SESSION V:
BLUEBERRY NUTRITION
Moderator: Jim Spiers

Organic Matter and Nitrogen Level Effects on Mycorrhizal Infection in 'Bluecrop' Highbush Blueberry Plants

Barbara L. Goulart
Kathleen Demchak
Wei Qiang Yang

SUMMARY. A field planting of 'Bluecrop' highbush blueberry was established in spring of 1992 to evaluate the effects of various cultural practices on plant growth and mycorrhizal level. Treatments included mulch or no mulch, pre-plant amendment or no pre-plant amendment, and 4 levels of nitrogen fertilization (from 0 to 100 g/

Barbara L. Goulart is Associate Professor, Kathleen Demchak is Research Aide, and Wei Qiang Yang is a graduate student, Department of Horticulture, Penn State University, University Park, PA 16802.

[Haworth co-indexing entry note]: "Organic Matter and Nitrogen Level Effects on Mycorrhizal Infection in 'Bluecrop' Highbush Blueberry Plants." Goulart, Barbara L., Kathleen Demchak, and Wei Qiang Yang. Co-published simultaneously in *Journal of Small Fruit & Viticulture* (Food Products Press, an imprint of The Haworth Press, Inc.) Vol. 3, No. 4, 1995, pp. 151-164; and: *Blueberries: A Century of Research* (ed: Robert E. Gough, and Ronald F. Korcak) Food Products Press, an imprint of The Haworth Press, Inc., 1995, pp. 151-164. Single or multiple copies of this article are available from The Haworth Document Delivery Service [1-800-342-9678, 9:00 a.m. - 5:00 p.m. (EST)].

151

plant in year 2, and from 0-120 g/plant in year 3) arranged in a complete factorial experiment. After two years, interactions among the treatments characterized the plant's responses. When no mulch was employed, increasing nitrogen level resulted in decreased mycorrhizal concentration in the roots. When plants were mulched, effects were inconsistent. Mulch and/or amendment increased plant growth and vigor. For plants with no mulch and no amendment, canopy volume and mycorrhizal concentration level decreased with increasing nitrogen. *[Article copies available from The Haworth Document Delivery Service: 1-800-342-9678.]*

KEYWORDS: Highbush blueberry; *Vaccinium corymbosum*; Nutrition; Nitrogen; Fertilization; Symbiosis; Ericoid mycorrhizae; Organic matter; Mulches; Soil amendment; *Hymenoscyphus ericae*

INTRODUCTION

Highbush blueberry roots, like those of most ericaceous plants, host a fungal symbiont which increases the plant's ability to forage for nutrients, most notably nitrogen. The fungi benefit from this relationship because it utilizes the host plant's photosynthates as its carbohydrate source. The fungal symbiont, usually an ascomycete for ericaceous plants, infects the root and forms mycorrhizae (which literally means "fungus roots"). While many studies have been conducted on the effects of organic matter (both pre-plant and as a mulch) and nitrogen form and level on blueberries, and some studies have been conducted on the effect or level of mycorrhizal infection on blueberries, none of these studies have reported the relationship between nutrition and planting practices with the level of mycorrhizal infection.

Organic matter and nitrogen studies. Soil amendments for pre-plant incorporation improve blueberry establishment and production on mineral soils. Perlmutter and Darrow (1942) found that blueberry seedlings grew best in a media that was 2/3 "forest litter" and 1/3 soil. More recently, Odneal and Kaps (1990) found that aged or fresh pine bark was as good as sphagnum peat for preplant incorporation, with no differences in plant height, spread, number of canes or plant yield or size. Rotted sawdust in the planting hole has also proven to be a good material for preplant incorporation (Eck, 1988).

Mulches are considered essential for blueberry production on mineral soils. Numerous studies have been undertaken to evaluate the effect of mulches. On rabbiteye blueberries grown on a fine sandy loam, mulching increased plant height, and shoot and root weights (Patten et al., 1988). Many researchers have examined different types of mulches for their effectiveness. Savage (1942) found that 8 cm (3 in) of sawdust mulch (at least 1 year old) was superior to rye straw or oak leaf mulches, with better plant survival and growth. All mulches, however, were notably better than clean cultivation. Skirvin and Otterbacher (1984, 1986, 1987) and Meador et al. (1984) have found that many mulching materials work well. Sawdust and chopped corn stalks were particularly effective. Leaf mulch (of oak, locust, beech, maple, sycamore and some pine needles) also performed as well as sawdust and chopped corn stalks, but had to be replenished yearly rather than once every two years. All mulches were aged 1-2 years. Wood chips and straw produced the lowest yields and shortest plants. Pine bark and well composted manure or stable bedding have been reported to work well (Eck, 1988), however, the latter materials may increase soil pH. Korcak (1988) cites the reasons for improved performance using organic mulches as improved water-holding capacity, reduced temperature fluctuations, weed control and improved soil tilth. Mulches also maintain a more constant media pH, and in some cases contribute to acidification of the soil.

A study by Spiers (1986) found that on rabbiteye blueberries, mulch was the most important component, followed by incorporated peatmoss and irrigation. Mulching resulted in a uniform root distribution from the plant crown outward, with most roots in the upper 15 cm (6 in) of soil. Incorporated peatmoss tended to concentrate the root system near the crown area (most at the 30-45 cm [12 to 18 in] depth). After three years, all of the plants that had received none of the treatments (no mulch, peat moss or irrigation) had died.

Recommended levels of nitrogen fertilizer application for mineral soils are between 45 and 91 kg of actual nitrogen/ha (40 and 82 lbs of actual nitrogen/A) (urea or ammonium sulfate) depending upon plant age (Goulart et al., 1994). There has been considerable debate regarding the nitrogen form utilized by highbush blueberry plants (Townsend, 1967; Hammett and Ballinger, 1972; Rosen et

al., 1990). Most studies conclude that there is preferential utilization of ammonium nitrogen by *Vaccinium*, making it an unusual
species, since most plants prefer to take up nitrate nitrogen (Peterson et al., 1988; Townsend, 1967; Merhaut and Darnell, unpublished). Studies on *Calluna vulgaris* and *Vaccinium macrocarpon*
resulted in increased total N, increased N concentration in the shoot,
and increased shoot growth on mycorrhizal versus non-mycorrhizal
plants (Read and Stribley, 1973; Stribley and Read, 1974). Mycorrhizal plants also contained more total nitrogen than non-mycorrhizal plants, suggesting that the mycorrhizal plants were better able to
utilize non-ammonium sources of nitrogen than non-mycorrhizal
plants (Stribley and Read, 1974). Other research demonstrated that
ericoid mycorrhizae can assimilate amino acids, peptides and proteins from the soil, and transfer N from these sources to host plants,
providing an additional source of N which plants otherwise would
not be able to access (Bajwa and Read, 1986).

Mycorrhizal studies on Vaccinium. Results on the influence of
mycorrhizae on *V. corymbosum* growth and development are inconsistent. Although Reich et al. (1982) reported no significant differences in growth or shoot number on blueberries when plants were
inoculated with two mycorrhizal isolates, control plants were as
infected as inoculated plants. Powell and Bates (1981) reported
increased yield on 6 blueberry cultivars when they were planted
with peat that had been dug from a highly infected natural population of blueberries. In a later study, blueberries inoculated with
mycorrhizal peat were more highly mycorrhizal than plants inoculated with a mycelial suspension of *H. ericae*, suggesting that the
source of inoculum in the peat is more infective than the more often
used *H. ericae* (Powell and Bagyaraj, 1984). *V. angustifolium* has
also been shown to support mycorrhizal infection under controlled
conditions (Smagula and Litten, 1989).

Surveys of mycorrhizal infection intensity in native and commercial populations of *V. corymbosum* have also yielded inconsistent
results. Survey work in the southeastern United States found low
levels of infection (1-5%) on young commercial plantings of *V.
ashei*, and greater colonization (about 50%) on older plantings (Jacobs et al., 1982). Surveys of *V. corymbosum* in North Carolina
found only 1-3% mycorrhizal infection in commercial plantings,

with higher levels (up to 85%) of infection in native populations (Boyer et al., 1982). To our knowledge, native populations of *V. angustifolium* or *V. darrowi* have not been screened for the presence of mycorrhizal infection. A more recent survey of native and commercial populations of *Vaccinium* in Pennsylvania, New Jersey and Michigan found varying levels of infection, but the levels were, in general, much higher than previously reported (Goulart et al., 1993).

The objective of this study was to evaluate the effects of nitrogen fertilization and organic amendment practices on level of mycorrhizal infection in field-grown highbush blueberry.

MATERIALS AND METHODS

Two-year old bareroot 'Bluecrop' highbush blueberry plants were established at the Horticulture Farm of the Russell E. Larsen Agricultural Research Center in Rock Springs, Pennsylvania in a Hagerstown silt loam soil in May of 1992. Plants had 50% of their bearing surface removed at planting. Plants were watered with trickle irrigation as needed (at least 5 cm [2 in] of water/week), and pests controlled using local recommendations (Goulart et al., 1991). Nitrogen was not applied in the planting year.

Plants were subjected to the following treatments in a completely crossed factorial experiment:

1. Pre-plant soil amendment (none, or 8 liters [2 gal] of rotted sawdust worked into the planting hole),
2. Mulch or no mulch, and
3. Nitrogen level (in 1993: 0, 33, 66 or 100 g [0, 1.15, 2.30, or 3.45 oz] ammonium sulfate/plant; in 1994: 0, 40, 80 and 120 g [0, 1.4, 2.8, or 4.2 oz] ammonium sulfate/plant).

Each treatment unit consisted of 3 plants, and was replicated 8 times in a randomized complete block design. Data was collected from the center plant of each plot, and was analyzed using an analysis of variance, and where appropriate, means separated using Tukey-Kramer's mean separation test. For mycorrhizal infection level analysis, square root transformations of the original percentages were analyzed.

The percent of the roots which were infected with mycorrhizae was evaluated every fall and spring from fall 1992 through spring 1994. Root samples were cleared and stained according to the technique described by Koske and Gemma (1989) and percentage infection quantified used by the grid-line technique developed by Giovannetti and Mosse (1980). Plant survival was also recorded during spring of 1993 and 1994. On 24 September 1993, photosynthesis rate (A) and associated parameters were recorded using an ADC LCA-2 battery-powered portable infrared CO_2 analyzer. Plant water use efficiency (WUE) was calculated by dividing A by transpiration rate. Plant height was also recorded in Fall 1993, as well as canopy volume (calculated by measuring the widest canopy diameter and the diameter perpendicular to the first measurement, averaging the two widths, and multiplying the area of the circle formed by the average diameter by plant height). Foliage density was also evaluated subjectively at this time (0, 20, 40, 60, 80 or 100 percent coverage of the plant limbs), and an index of plant vigor developed. This index was the canopy volume × the foliage density, and will be referred to hereafter as the canopy index.

RESULTS AND DISCUSSION

No treatment had any effect on plant survival (overwintering) in Spring, 1993 (dns). However, by the following year, 9 plants had died, and all were in the no mulch, no amendment treatments.

Plant mycorrhizal infection levels fluctuated relative to treatment over the 2 years of the study, indicating that the level of infection in the roots is rather dynamic (Table 1). In Fall of 1992, unamended, unmulched plants had higher levels of infection than unamended mulched plants, however, among the amended treatments, the presence or absence of mulch did not affect infection level. It should be noted that this was a particularly wet season, and the soil around the plants which were amended and mulched was saturated, whereas that around the unamended, unmulched treatments was relatively dry. No differences in level of infection were detected in spring of 1993; however, by fall of 1993, mulched plants had higher mycorrhizal levels, regardless of amendment treatment. The summer of 1993 was particularly hot, and moderately dry, though plots were all

TABLE 1. Effect of amendment and mulch treatments on level of mycorrhizal infection in field-grown 'Bluecrop' highbush blueberry.

Treatment	Fall 1992	Spring 1993	Fall 1993	Spring 1994
Amended/Mulched	2.6	3.9	0.9	2.2
Amended/Unmulched	3.0	4.4	6.5	10.0
p(F)	0.33	0.66	< 0.001	< 0.001
Unamended/Mulched	2.8	6.0	3.2	8.5
Unamended/Unmulched	7.7	6.6	6.7	6.8
p(F)	< 0.001	0.90	0.02	0.81

trickle irrigated as necessary. By the spring of 1994, there was an interaction between amendment and mulch treatments. Specifically, mulched amended plants had low levels of infection compared to unmulched amended plants, while those which were unamended were unaffected by mulch. Further interactions with nitrogen treatment are discussed later in this paper.

Vegetative parameters and their relationship to mycorrhizal infection level. In Fall of 1993, there were no statistical interactions between mulch and amendment treatments. Mulched plants were taller, had larger canopy volume and canopy index values than unmulched plants (Table 2). There were no differences in A or WUE, and mycorrhizal infection levels were significantly higher on unmulched versus mulched plants. This reduction of mycorrhizal infection in mulched plants was unexpected, and may be due more to an increase in root volume than a real decrease in actual number of infection sites. A previous greenhouse study showed that increasing nitrogen decreased mycorrhizal concentration due to a dilution effect from increased root growth (Goulart et al., 1993).

Plants which had been amended were also taller, had larger canopy volume, higher canopy indices and higher A and WUE than unamended plants. Level of mycorrhizal infection was unaffected by amendment treatment.

By the Spring of 1994, interactions among mulch, amendment and nitrogen level precluded analysis by each parameter indepen-

TABLE 2. The effect of mulch and pre-plant amendment on vegetative characteristics of field-grown 'Bluecrop' highbush blueberry: Fall, 1993.

Treatment[z]	Height (cm)	Canopy[y] Volume (cm³)	Foliage[x] Density	Canopy[w] Index	A[v]	WUE[u]	Mycorrhizae % Infection
Mulched	51.4	61177	3.8	49146	8.1	11.3	2.3
Unmulched	44.3	34743	2.9	22018	7.8	10.8	7.6
p(F)	< 0.001	< 0.001	< 0.001	< 0.001	0.36	0.44	< 0.001
Amended	50.8	55139	3.4	41764	9.4	13.1	4.1
Unamended	44.8	39925	3.3	31058	6.4	8.8	5.5
p(F)	0.01	0.03	0.49	0.03	< 0.001	< 0.001	0.33

[z] Treatment means analyzed using Analysis of Variance.
[y] Canopy volume calculated by measuring the widest diameter of the canopy and the diameter perpendicular to the widest diameter, and calculating the volume of the cylinder formed by the average of these two diameters and the plant height.
[x] Foliage density: 1 = sparse (about 20% of the canopy area filled with leaves) thru 5 = dense (about 100% of the canopy area filled with leaves).
[w] Canopy index = canopy volume*foliage density.
[v] A = rate of photosynthesis in μeinsteins/M²/sec.
[u] WUE = water use efficiency, calculated as A/transpiration rate.

dently (Table 3). Among amended plants, those which were mulched had higher foliage densities and higher canopy indices than those which were not mulched. These plants also had much lower levels of mycorrhizal infection. In other words, plants which appeared to be more vigorous had lower concentrations of mycorrhizae.

For plants which were not amended at planting, mulch had a more global effect, with taller plants, higher canopy volumes, higher foliage density ratings and higher canopy indices than for plants which were not mulched. This data concurs with findings by Spiers (1986) which found that mulch resulted in a more significant increase in plant growth than either preplant amendment or trickle irrigation on rabbiteye blueberries. Mycorrhizal level was unaffected by mulch treatment on unamended soils.

The combination of soil amendment and mulch clearly increased canopy volume and decreased level of mycorrhizal infection; however, there was also an interaction between mulch, amendment and

TABLE 3. The effect of mulch and pre-plant amendment on vegetative characteristics of field-grown 'Bluecrop' highbush blueberry: Spring, 1994.

Treatment[z]	Height (cm)	Canopy[y] Volume (cm³)	Foliage[x] Density	Canopy[w] Index	A[v]	T[u]	Mycorrhizae % Infection
Amended/							
Mulched	50.2	66722	4.4	59748	6.2	3.9	2.2
Amended/							
Unmulched	51.0	60312	3.6	47766	6.9	4.5	10.0
p(F)	0.59	0.36	< 0.03	< 0.08	0.53	0.01	< 0.001
Unamended/							
Mulched	47.2	63823	3.9	53466	6.3	4.1	8.5
Unamended/							
Unmulched	35.7	23946	3.1	16981	6.8	5.0	6.8
p(F)	< 0.001	< 0.001	0.001	< 0.001	0.97	0.25	0.81

[z] Treatment means analyzed using Analysis of Variance.
[y] Canopy volume calculated by measuring the widest diameter of the canopy and the diameter perpendicular to the widest diameter, and calculating the volume of the cylinder formed by the average of these two diameters and the plant height.
[x] Foliage density: 1 = sparse (about 20% of the canopy area filled with leaves) thru 5 = dense (about 100% of the canopy area filled with leaves.
[w] Canopy index = canopy volume*foliage density.
[v] A = rate of photosynthesis in µeinsteins/M²/sec.
[u] T = transpiration rate in mmoles/M²/sec.

nitrogen level (Figures 1 and 2). When plants were mulched, increasing levels of nitrogen resulted in increased canopy volume (Figure 1a, 1b), but had little effect on mycorrhizal level (Figure 2a, 2b). However, when plants were not mulched but amended, optimal nitrogen level was about 80 g/plant (Figure 1c). For plants with no mulch and no amendment, increasing nitrogen decreased canopy volume, and many plants died (Figure 1d). Mycorrhizal level decreased with increasing nitrogen in plants from both unmulched treatments (Figure 2c, 2d), though this negative relationship was strongest when amendment was incorporated into the planting hole,

FIGURE 1. Mulch, amendment and nitrogen effects on canopy volume of field grown 'Bluecrop' highbush blueberry. a. mulch, amendment ($y = 61511 - 11442x + 4508.8x^2$); b. mulch, no amendment ($y = 47154 - 21444x + 9370.5x^2$); c. no mulch, amendment ($y = 6729.4 + 47024x - 8529.7x^2$); d. no mulch, no amendment ($y = 93017 - 54711x + 8959.2x^2$).

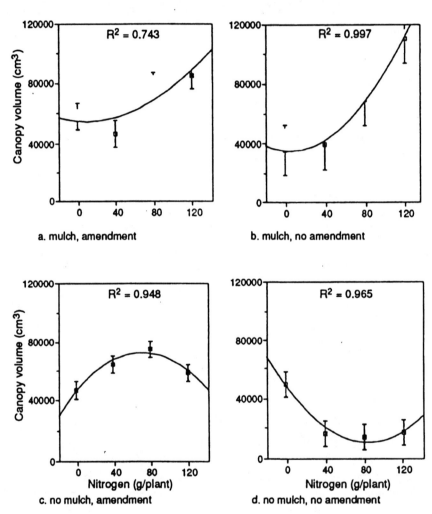

FIGURE 2. Mulch, amendment and nitrogen effects on percent mycorrhizal infection of field grown 'Bluecrop' highbush blueberry. a. mulch, amendment ($y = -2.5625 + 6.0208x - 1.3958x^2$); b. mulch, no amendment ($y = 7.25 + 0.21333x + 0.1x^2$); c. no mulch, amendment ($y = 12.522 + 2.9667x - 1.3456x^2$); d. no mulch, no amendment ($y = 13.271 - 2.9216x + 0.11341x^2$).

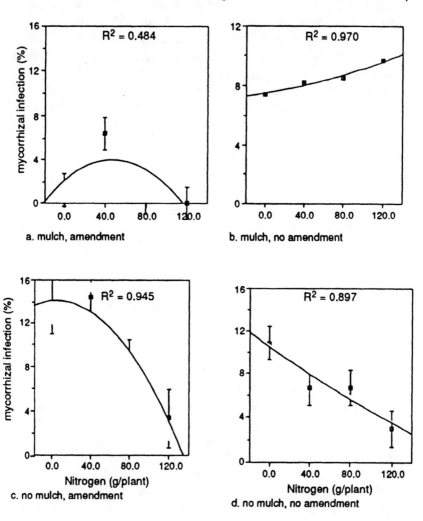

with an $r^2 = 0.945$. These data suggest that practices that increase blueberry plant growth and vigor do not increase the concentration of mycorrhizas in the roots. While the inverse relationship between mycorrhizal concentration and nitrogen level is predictable from the literature, the low levels associated with mulched amended plots, regardless of nitrogen level, was not. It should be noted that this data is preliminary in nature, and several more years worth of data will be collected and analyzed, and root systems excavated, before final conclusions are appropriate.

The effect of the treatments on mycorrhizal levels, and how those infection levels impact our cultural management, is unclear. While mycorrhizas offer the potential to reduce nitrogen usage and benefit the plant in other ways, our current methods of mulching and adding amendment to the planting hole generally seem to reduce the *intensity* of mycorrhizal infection. It will not be clear until the root systems are excavated in future years whether or not the absolute levels of infection (i.e., number of cells infected relative to the total dry weight of the root system) are enhanced or suppressed by these cultural management practices.

GROWER BENEFITS

The benefit of using mulch and pre-plant soil amendment, (in this case, rotted sawdust), in mineral soil plants was affirmed, with mulch showing the greatest benefit. Using the recommended level of nitrogen without mulch or amendment resulted in reduced plant growth and increased plant death, so that if mulch and amendment aren't used, a reduced rate of nitrogen is necessary. However, the results of this study make it abundantly clear that, for mineral soils such as these, mulch and amendment are a cultural necessity.

Mycorrhizas offer the potential to reduce fertilizer and pesticide inputs by taking advantage of the symbiotic relationship with respect to nutrient acquisition, avoidance of soil-related toxicity, and the amelioration of soil pathogen interactions. However, it is not clear how, or whether, we will be able to exploit this natural relationship to the improvement of agricultural systems or not.

LITERATURE CITED

Bajwa, R., and D.J. Read. 1986. Utilization of mineral and amino N sources by the Ericoid mycorrhizal endophyte *Hymenoscyphus ericae* and by mycorrhizal and non-mycorrhizal seedlings of *Vaccinium*. Trans. Br. Mycol. Soc. 87(2): 269-277.

Boyer, E.P., J.R. Ballington, and C.M. Mainland. 1982. Endomycorrhizae of *Vaccinium corymbosum*, L. in North Carolina. J. Amer. Soc. Hort. Sci. 107(5): 751-754.

Eck, Paul. *Blueberry Science.* 1988. Rutgers University Press, New Brunswick, New Jersey.

Giovannetti, M., and B. Mosse. 1980. An evaluation of techniques for measuring vesicular arbuscular mycorrhizal infection in roots. New Phytol. 84:489-500.

Goulart, B.L., K. Demchak, M.L. Schroeder, and J.R. Clark. 1993. Mycorrhizae in highbush blueberries: survey results and field response to nitrogen. HortScience 28(5):140.

Goulart, B.L., M. Brittingham, J. Harper, P. Heinemann, W. Hock, E. Rajotte, J. Rytter, and J. Travis. 1991. Small Fruit Production and Pest Management Guide, 1991-92. The Pennsylvania State University. 107 pp.

Goulart, B.L., M. Brittingham, J. Harper, P. Heinemann, W. Hock. 1994. Small Fruit Production and Pest Management Guide, 1994-95. The Pennsylvania State University. 115 pp.

Hammett, L., and W.E. Ballinger. 1972. A nutrient solution-sand culture system for studying the influence of N form on highbush blueberries. HortScience 7:498-499.

Jacobs, L.A., F.S. Davies, and J.M. Kimbrough. 1982. Mycorrhizal distribution in Florida Rabbiteye blueberries. HortScience 17(6):951-953.

Korcak, Ronald F. 1988. Nutrition of blueberry and other calcifuges. Horticultural Reviews 10:183-277.

Koske, R.E., and J.N. Gemma. 1989. A modified procedure for staining roots to detect VA mycorrhizas. Mycol. Res. 92(4):486-505.

Meador, D.B., C.C. Doll, J.W. Courter, R.M. Skirvin, and A.G. Otterbacher. 1984. Highbush blueberry cultural tips. Proc. 1984 Illinois Small fruit School. Horticulture Series 48:32-36.

Odneal, Marilyn B., and M.L. Kaps. 1990. Fresh and aged pine bark as soil amendments for establishment of highbush blueberry. HortScience 25:1228-1229.

Patten, K.D., E.W. Neuendorff, and S.C. Peters. 1988. Root distribution of 'Climax' rabbiteye blueberry as affected by mulch and irrigation geometry. Amer. Amer. Soc. Hort. Sci. 113:657-661.

Perlmutter, F., and G.M. Darrow. 1942. Effect of soil media, photoperiod and nitrogenous fertilizer on the growth of blueberry seedlings. Proc. Amer. Soc. Hort. Sci. 40:341-346.

Peterson, L.A., E.J. Stang, and M.N. Dana. 1988. Blueberry response to NH_4-N and NO_3-N. J. Amer. Soc. Hort. Sci. 113:9-12.

Powell, C.L., and D.J. Bagyaraj. 1984. Effect of mycorrhizal inoculation on the nursery production of blueberry cuttings–a note. New Zealand Jour. Agr. Res. 27:467-471.

Powell, C.L., and P.M. Bates. 1981. Ericoid mycorrhizas stimulate fruit yield of blueberry. HortScience 16(5):655-656.

Read, D.J., and D.P. Stribley. 1973. Effect of mycorrhizal infection on nitrogen and phosphorus nutrition of ericaceous plants. Nature New Biology 244:81-82.

Reich, L.A., R.F. Korcak, and A.H. Thompson. 1982. Effects of two mycorrhizal isolates on highbush blueberry at two soil pH levels. HortScience 17(4):642-644.

Rosen, C.J., D.L. Allan, and J.J. Luby. 1990. Nitrogen form and solution pH influence growth and nutrition of two *Vaccinium* clones. J. Amer. Soc. Hort. Sci. 115:83-89.

Savage, E.F. 1942. Growth response of blueberries under clean cultivation and various kinds of mulch materials. Proc. Am. Soc. Hort. Sci. 40:335-337.

Skirvin, R.M., and A.G. Otterbacher. 1984. The effect of various mulch materials on the yield of young blueberry plants. Proc. 1984 Illinois Small Fruit School. Horticulture Series 48:38-39.

Skirvin, R.M., and A.G. Otterbacher. 1986. The effect of various mulch materials on the yield of young blueberry plants. Proc. 1984 Illinois Small Fruit School. Horticulture Series 57:15-16.

Skirvin, R.M., and A.G. Otterbacher. 1987. The effect of various mulch materials on the yield of young blueberry plants. Proc. 1984 Illinois Small Fruit School. Horticulture Series 63:20-21.

Smagula, J.M., and W. Litten. 1989. Effect of ericoid mycorrhizae isolates on growth and development of lowbush blueberry tissue culture plantlets. Acta Horticulturae 241:110-112.

Spiers, James M. 1986. Root distribution of 'Tifblue' rabbiteye blueberry as influenced by irrigation, incorporated peatmoss, and mulch. J. Amer. Soc. Hort. Sci. 111:877-80.

Stribley, D.P., and D.J. Read. 1974. The biology of mycorrhiza in the Ericaceae. New Phytol. 73:1149-1155.

Stribley, D.P., and D.J. Read. 1976. The biology of mycorrhiza in the ericaceae. VI. The effects of mycorrhizal infection and concentration of ammonium nitrogen on growth of cranberry (*Vaccinium macrocarpon*, Ait.) in sand culture. New Phytol. 77:63-72.

Townsend, L.R. 1967. Effect of ammonium nitrogen and nitrate nitrogen, separately and in combination, on the growth of the highbush blueberry. Can. J. Plant Sci. 49:333-338.

Blueberry Nitrate Reductase Activity:
Effect of Nutrient Ions and Cytokinins

Shiow Y. Wang
Ronald F. Korcak

SUMMARY. Nitrate Reductase activity (NRA) was detected in the blueberry cultivars *(Vaccinium* sp.), 'Bluecrop' and 'Tifblue.' Shoots treated with KNO_3 induced the highest level of NRA in leaves and stems of both cultivars. In 'Bluecrop' and 'Tifblue,' the older leaves showed higher NRA compared to young leaves. Aluminum and sulfate ions and various forms of NH_4^+ showed an inhibitory effect on NRA in both cultivars. NRA in younger leaves was more sensitive to the addition of various ions than in the older leaves, and 'Bluecrop' NRA was more sensitive than 'Tifblue' NRA. Cytokinins (kinetin, zeatin, zeatin riboside and 6-benzylaminopurine) supplemented with nitrate showed an additive stimulation of NRA under light and dark. However, under light, the level of NRA in blueberry induced with cytokinins was much greater than that obtained in the dark. *[Article copies available from The Haworth Document Delivery Service: 1-800-342-9678.]*

INTRODUCTION

Blueberry *(Vaccinium* sp.) is an economically-important calcifuge (lime-avoiding plant). There have been many contradictions in

Shiow Y. Wang is Plant Physiologist/Biochemist, and Ronald F. Korcak is Soil Scientist, Fruit Laboratory, Beltsville Agricultural Research Center, Agriculture Research Service, U.S. Department of Agriculture, Beltsville, MD 20705.
The authors wish to thank Sue J. Mok for technical assistance.

[Haworth co-indexing entry note]: "Blueberry Nitrate Reductase Activity: Effect of Nutrient Ions and Cytokinins." Wang, Shiow Y., and Ronald F. Korcak. Co-published simultaneously in *Journal of Small Fruit & Viticulture* (Food Products Press, an imprint of The Haworth Press, Inc.) Vol. 3, No. 4, 1995, pp. 165-181; and: *Blueberries: A Century of Research* (ed: Robert E. Gough, and Ronald F. Korcak) Food Products Press, an imprint of The Haworth Press, Inc., 1995, pp. 165-181. Single or multiple copies of this article are available from The Haworth Document Delivery Service [1-800-342-9678, 9:00 a.m. - 5:00 p.m. (EST)].

the literature concerning the source of N for its growth and development. Many investigators have indicated NH_4-N was a better N source than NO_3-N (Cain, 1952; Townsend, 1966; 1967), while others (Dirr et al., 1972; Holmes, 1960; Oertli, 1963) have stated that N source made little difference on blueberry growth. Nitrate reductase (NR) is a key enzyme in the nitrogen cycle (Shen et al., 1993). Nitrogen is generally transported in plants in a reduced form (Pate, 1980; Titus and Kang, 1982). This suggests that most reduction takes place in the roots (Lee and Titus, 1992). However, recent research has proven that nitrate reductase activity (NRA) can be found in leaves and stems in various fruit plants (Lee and Titus, 1992). The amount of reduced nitrogen is correlated to the level of NRA (Muller and Janiesch, 1993). NRA level is a good indicator of the rate of nitrogen metabolism (Shen et al., 1993). The rate of NRA is closely associated with nitrate assimilation, plant growth and ultimately to crop yield (Shen et al., 1993). NRA levels are also related to growth rate and the availability of metabolites (Wakhloo and Staudt, 1988). NRA in the tissue may be regulated by exogenous nitrate and other nutrient elements. Cytokinins and light have been known to affect NR activity (Kapoor et al., 1987; Kuznetsov et al., 1979; Lu et al., 1983). The purpose of this study was to determine the effect of various nutrient sources and cytokinin treatments under light and dark conditions on NRA in leaves and stems of two different cultivars of blueberry, 'Bluecrop' (*V. corymbosum)* and 'Tifblue' (*V. ashei*).

MATERIALS AND METHODS

Plant Cultivation

Blueberry branches (approximately 14-cm long) were collected from the orchard at the Beltsville Agricultural Research Center and raised in various solutions to allow uptake of nutrients. Shoots in H_2O were used as the control. Experimental shoots were allowed to uptake either KNO_3, NH_4NO_3, or different types of nutrients. Samples were taken after 24, 48, or 96 hr to assay NRA. To determine the effect of various cytokinins on NRA of blueberry, shoots of

'Bluecrop' or 'Tifblue' were dipped daily for 5 min in various concentrations of kinetin (1, 5, 10, and 100 μM) or various types of cytokinins. Tween-20 was added in a 0.5% (v/v) to improve absorption. Control shoots were dipped in water. All shoots were raised in 5 mM KNO_3 in light versus dark until assays were performed. NRA was determined after 72 hrs.

NRA Assay

The *in vivo* NR assay was based on the assays of Hageman and Reed (1980). Leaves from various treatments were cut into discs with a 1 cm diameter (using a #4 cork borer). Stem samples were collected and cut into 1 cm long pieces. Approximately 0.5 g of leaf discs or stems were weighed into test tubes. Five ml of assay solution containing 100 mM potassium phosphate buffer (pH 7.5) and 2% n-propanol was added to each tube. Thirty mM potassium nitrate was also added to the assay medium. The samples were then vacuum infiltrated (5 mm Hg) twice for 3 min and incubated in the dark in a shaking water bath for 60 min at 30°C. The amount of nitrite produced was determined as a measure of NRA by developing the azo color complex with 1% sulfanilamide in 3 N HCl and 0.12% N-(1-naphthyl-ethylene diamine). After allowing 30 min for full color development, absorbance was determined at 540 nm using a spectrophotometer (Shimadzu UV-160A, Columbia, MD). A standard curve in the range 0-100 μmol KNO_2 was used. The enzyme activity was expressed as nmoles NO_2 produced/hr/g fresh weight.

RESULTS

Effect of Nitrate Supply on NRA

Nitrate reductase activity was detected in leaves and stems of blueberry in response to various forms of nitrogen (Figures 1-3). All the shoots appeared healthy with no chlorosis or toxicity symptoms when NH_4-N or NO_3-N was provided. Nitrate supply in the solution significantly increased the NRA in leaves and stems of both blueberry cultivars (Figures 1-3). Extending the time during which the shoots were in the nitrate solution also increased the NRA (Figures 1-3). There appears to be a higher amount of induced NR activity in leaves of both cultivars when KNO_3 was added versus NH_4NO_3. Tifblue blueberry leaves showed a significant increase in NRA when the plant was raised in 5 mM KNO_3 or 5 mM NH_4NO_3 versus H_2O (Figure 1). In 'Tifblue' stems, the NR activity level induced by NO_3 and NH_4NO_3 was approximately equal (Figure 3). When comparing these results to another cultivar, Bluecrop, the general pattern varied only slightly. 'Bluecrop' leaves responded much better to KNO_3 than NH_4NO_3 although both induced NRA significantly more than H_2O (Figure 2). In stems of 'Bluecrop,' the results were very similar with an even greater difference between the induced enzyme activity of the KNO_3 treated and NH_4NO_3 treated shoots (Figure 3). Shoots treated with KNO_3 induced the highest level of NRA in all samples of both cultivars, while NH_4NO_3 induced greater levels than H_2O.

Effect of Various Forms of NH_4^+ on NO_3^- Induced NRA

With the addition of various forms of NH_4^+ ions on NO_3^- induced NRA, a general trend of inhibition was observed (Tables 1 and 2). The differences between the sensitivity of the two cultivars towards the treatments were clearly visible. 'Tifblue' seemed to be less affected by the various forms of NH_4^+, whereas 'Bluecrop' NRA was highly sensitive toward the treatments. Nitrate reductase activity was measured after 48 hr or 72 hr uptake of various NH_4^+ solutions. Shoots treated with 10 mM KNO_3 (5 mM KNO_3 plus 5 mM KNO_3) were used as the control and designated as 100% NRA activity. All other NRA was measured in comparison to KNO_3. When an additional 5 mM NH_4NO_3 was added, a decrease in NRA was clearly observed in 'Bluecrop.' However, 'Tifblue' is a more resistant cultivar and is less affected by NH_4^+ ions. 'Bluecrop' leaf NRA responded to the added NH_4NO_3 by dropping to about 80% of the control, whereas 'Tifblue' only decreased to 97% of the control (Tables 1 and 2). In stems, the results were very similar to those obtained with leaf samples (data not shown).

NH_4Cl inhibited NRA levels to some degree in both blueberry cultivars. NRA levels in 'Bluecrop' were decreased significantly

FIGURE 1. Effect of NO_3^- and NH_4^+ on the nitrate reductase activity (NRA) in leaves of 'Tifblue' (TB) blueberry. 'Tifblue' shoots were exposed to H_2O, 5 mM KNO_3, or 5 mM NH_4NO_3 and NRA was determined after 24, 48, 72, and 96 hr. The results are means of three replications ± SE.

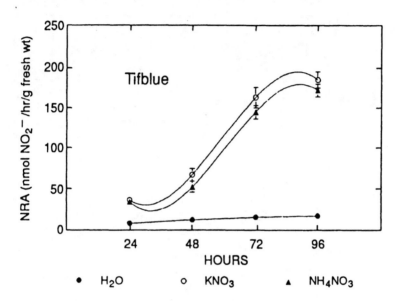

more than in 'Tifblue,' indicating a higher sensitivity to NH_4Cl in 'Bluecrop.' Using randomly selected leaves, NRA decreased to 69.7% of control in 'Bluecrop' while it only decreased to 88.5% of the control in 'Tifblue' (Table 1). Assays of the upper, younger leaves of 'Bluecrop' appeared to be more sensitive, decreasing to 45.7% of control whereas the older leaves were less inhibited (85.3% of control) (Table 2). 'Tifblue' did not vary in sensitivity between upper and lower leaves. Differences between cultivars were evident (Table 2). However, 'Tifblue' seems to be less inhibited by the addition of NH_4Cl (Tables 1 and 2).

Acetate ions produced effects in both cultivars NRA similar to those of NH_4Cl. When formate or oxalate was provided to the two cultivars, a decrease in NRA was found (Table 1). When sulfate was provided to the shoots, results were as expected with a decrease in NRA in the younger leaves which were more sensitive when compared to older leaves (Tables 1 and 2). Toxicity was clearly visible

FIGURE 2. Effect of NO_3^- and NH_4^+ on the nitrate reductase activity (NRA) in leaves of 'Bluecrop' (BC) blueberry. 'Bluecrop' shoots were exposed to H_2O, 5 mM KNO_3, or 5 mM NH_4NO_3 and NRA was determined after 24, 48, 72, and 96 hr. The results are means of three replications ± SE.

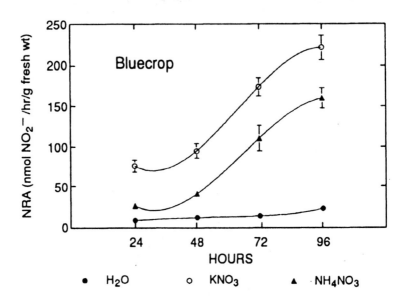

in shoots treated with sulfamate. Leaves appeared to be dying and the level of NRA was extremely low (Table 1). 'Bluecrop' showed a greater sensitivity to added sulfamate ion dropping to 15.8% of the control, while 'Tifblue' dropped to 19.5% (Table 1). The cultivars were also exposed to $AlCl_3$ and the effect on NRA was approximately the same, with both cultivars, at approximately 30% of the control (Table 1). In general, NRA was greatest in the leaves and stems in NO_3-N solution for both cultivars. Other forms of NH_4^+ showed an inhibition on NRA, with 'Bluecrop' being more sensitive compared to 'Tifblue' in response to various forms of NH_4^+ ions.

Effect of Various Forms of SO_4^{2-} Ion on NO_3^- Induced NRA

Exposing blueberry shoots to H_2O, 5 mM KNO_3, or 5 mM NH_4NO_3 plus various forms of SO_4^{2-} also had significant effects on NRA (Table 3). When $FeSO_4$, $MgSO_4$, or $Al_2(SO_4)_3$ was added,

FIGURE 3. Effect of NO_3^- and NH_4^+ on the nitrate reductase activity (NRA) in stems of 'Bluecrop' (BC) and 'Tifblue' (TB) blueberry. BC and TB shoots were exposed to H_2O, 5 mM KNO_3, or 5 mM NH_4NO_3 and NRA was determined after 96 hr. The results are means of three replications ± SE.

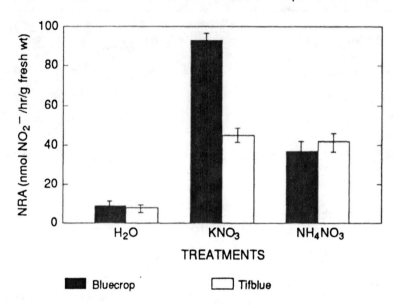

the NRA level decreased in all samples, whereas the effects of $MnSO_4$ and $CaSO_4$ varied slightly between the cultivars. 'Bluecrop' NRA level decreased to approximately 70% of the control, while 'Tifblue' NRA level increased to nearly 35% of the control (Table 3).

Effect of Cytokinins on NO_3^- Induced NRA

When kinetin was supplemented with nitrate, an additive effect was observed, and the efficiency of NRA induction in leaves and stems markedly increased in both cultivars (Tables 4 and 5). Kinetin showed concentration-dependent stimulation, and was inhibitory at higher concentrations (100 µM). The maximum increase in NRA was obtained at the 10 µM concentration of kinetin (Tables 4 and 5). The effects of kinetin on NRA in 'Bluecrop' and 'Tifblue' were comparable (Tables 4 and 5).

The effect of other cytokinins (zeatin, zeatin riboside, and 6-ben-

TABLE 1. Nitrate reductase activity (NRA) of leaves of 'Bluecrop' and 'Tifblue' blueberry. Shoots were exposed to 5 mM KNO_3 plus 5 mM of various forms of NH_4^+ ion or $AlCl_3$. NRA activity was determined after 48 hr. The results are means of three replications ± SE.

Treatment	'Bluecrop'	% control	'Tifblue'	% control
	NRA (nmol NO_2^-/hr/g fresh wt)			
KNO_3 + KNO_3	165 ± 12	100.0	113 ± 11	100.0
KNO_3 + NH_4NO_3	131 ± 12	79.4	110 ± 10	97.3
KNO_3 + NH_4Cl	115 ± 9	69.7	100 ± 8	88.5
KNO_3 + $CH_3CO_2NH_4$ (acetate)	105 ± 11	63.6	76 ± 8	67.3
KNO_3 + HCO_2NH_4 (formate)	83 ± 6	50.3	75 ± 4	66.4
KNO_3 + $(NH_4)_2C_2O_4 \cdot H_2O$ (oxalate)	81 ± 10	49.1	69 ± 9	61.1
KNO_3 + $(NH_4)_2SO_4$	63 ± 7	38.2	55 ± 7	48.7
KNO_3 + $H_2NSO_3NH_4$ (sulfamate)	26 ± 3	15.8	22 ± 2	19.5
KNO_3 + $AlCl_3$	56 ± 2	33.9	36 ± 3	31.9

zylaminopurine) was also examined in 'Bluecrop' leaves and stems (Tables 6 and 7). Each cytokinin, when applied with nitrate either under light or dark, stimulated NRA. However, in the presence of light, the levels of induction were more than those observed in the dark (Tables 4-7). The cytokinins that induced the highest levels of NRA among the four cytokinins tested was zeatin. This was followed by kinetin, zeatin riboside, and 6-benzylaminopurine. However, the results obtained were very similar with zeatin riboside, and 6-benzylaminopurine.

TABLE 2. Effect of various forms of NH_4^+ ion on NO_3^- induced nitrate reductase activity (NRA) in upper and lower leaves of 'Bluecrop' or 'Tifblue' blueberry. 'Bluecrop' or 'Tifblue' shoots were exposed to 5 mM KNO_3 plus 5 mM of various forms of NH_4^+ ion. NRA was determined after 72 hr. The results are means of three replications ± SE.

	NRA (nmol NO_2^-/hr/g fresh wt)			
	Upper leaves		Lower leaves	
Treatment	Bluecrop	Tifblue	Bluecrop	Tifblue
KNO_3 + KNO_3	175 ± 12	97 ± 4	184 ± 11	157 ± 14
KNO_3 + NH_4NO_3	130 ± 11	92 ± 6	160 ± 11	149 ± 14
KNO_3 + NH_4Cl	80 ± 4	85 ± 5	157 ± 9	141 ± 9
KNO_3 + $(NH_4)_2SO_4$	65 ± 4	35 ± 4	89 ± 6	86 ± 6

DISCUSSION

The induction of NRA levels has been shown to vary when shoots were allowed to uptake a number of different ions although the mechanism of ion action is not clearly understood. NRA is related to the availability of metabolites such as NO_3^-, NH_4^+, and K^+ (Wakhloo and Staudt, 1988). Thus far it appears that providing KNO_3 induces the greatest level of NRA in both cultivars. NRA has also been reported to be greatest in the leaves of highbush blueberry (*Vaccinium corymbosum* L.) grown on NO_3-N (Dirr et al., 1972). Increasing the time of exposure for shoots in nitrate solution also increased the NRA, reflecting the potential NRA when nitrate is not limiting.

Nitrate reductase activity was detected in stems and leaves of blueberry when shoots were allowed to uptake either H_2O, 5 mM KNO_3 or 5 mM NH_4NO_3. The shoots treated with 5 mM KNO_3 clearly showed the greatest induction of NRA and NH_4^+ was slightly less. The greater response to KNO_3 may be due in part to the K^+

TABLE 3. Effect of various SO_4^{2-} associated cations on NO_3^- induced nitrate reductase activity (NRA) in leaves of 'Bluecrop' (BC) or 'Tifblue' (TB) blueberry. Blueberry shoots were exposed to either H_2O, 5 mM KNO_3, or 5 mM NH_4NO_3 plus various forms of SO_4^{2-}. NRA was determined after 60 hr. The results are means of three replications $\pm$ SE.

		NRA (nmol NO_2^-/hr/g fresh wt)		
Treatment	Cultivars	H_2O	KNO_3	NH_4NO_3
Control	BC	30 ± 3	126 ± 11	91 ± 8
	TB	42 ± 4	104 ± 9	94 ± 6
$FeSO_4$	BC	21 ± 2	69 ± 4	61 ± 4
	TB	17 ± 2	78 ± 2	73 ± 2
$MgSO_4$	BC	28 ± 2	55 ± 2	57 ± 4
	TB	27 ± 2	80 ± 5	85 ± 6
$Al_2(SO_4)_3$	BC	20 ± 2	59 ± 4	54 ± 3
	TB	24 ± 2	64 ± 2	64 ± 4
$MnSO_4$	BC	30 ± 2	90 ± 6	84 ± 7
	TB	18 ± 2	105 ± 10	113 ± 10
$CaSO_4$	BC	27 ± 3	91 ± 8	83 ± 4
	TB	31 ± 3	139 ± 10	123 ± 11

concentration. In older, mature leaves of *Nicotiana tabacum,* an increase in K^+ concentration promoted the polymerization of primary metabolites (Wakhloo, 1985). It appears that K^+ concentration influences the availability of organic substrates which are needed for the synthesis of nitrate reductase (Wakhloo and Staudt, 1988).

Although reasons for the slight inhibitory effect by NH_4^+ compared to NO_3^- is not totally understood, perhaps NO_3-N served as a better inducer for NRA. Ammonium has been shown to be involved with energy conservation; less energy is necessary to assimilate ammonium than nitrate (Beck and Renner, 1989). It is not yet understood whether ammonium directly effects NRA or whether it contributes to the establishment of a larger metabolic

TABLE 4. Effect of kinetin on the nitrate reductase activity (NRA) in leaves of 'Bluecrop' and 'Tifblue' blueberry. 'Bluecrop' and 'Tifblue' shoots were dipped in various concentrations of kinetin solution for 5 min daily and exposed to 5 mM KNO_3. NRA was determined after 72 hr in light vs. dark. The results are means of three replications ± SE.

| | NRA (nmol NO_2^-/hr/g fresh wt) | | | |
| | Light | | Dark | |
Kinetin Concn (μM)	Bluecrop	Tifblue	Bluecrop	Tifblue
0	191 ± 10	130 ± 9	54 ± 3	44 ± 1
1	258 ± 10	210 ± 8	77 ± 5	51 ± 5
5	284 ± 12	236 ± 12	88 ± 4	59 ± 8
10	368 ± 12	267 ± 11	97 ± 5	68 ± 4
100	164 ± 8	114 ± 8	55 ± 2	40 ± 3

pool of nitrate (Ferrari et al., 1973). In general, the two cultivars of blueberry showed similar responses to these treatments. 'Tifblue' is less sensitive to the ions examined and more adaptable to adverse conditions compared to 'Bluecrop.'

In 'Bluecrop' and 'Tifblue,' mature leaves also showed higher NRA compared to young leaves. This is contradictory to findings by Roth-Bejerano and Lips (1970) who stated that NRA activity decreases with leaf age.

The effect of sulfate associated cations on NRA in blueberry was also examined. Evidence has suggested that the deprivation of sulfate depresses the transport and net influx of nitrate and ammonium due to some form of co-regulation (Clarkson et al., 1989). However, in this study, the addition of $MnSO_4$ and $CaSO_4$ promoted NRA in 'Tifblue,' whereas $FeSO_4$, $MgSO_4$ and $Al_2(SO_4)_3$ inhibited NRA in both cultivars. These appear to be counter-ion effects. The addition of Fe^{2+}, Mg^{2+} and Al^{3+} may be responsible for this inhibition. Although no literature is available on Fe^{2+} and Mg^{2+}, evidence has

TABLE 5. Effect of kinetin on the nitrate reductase activity (NRA) in stems of 'Bluecrop' and 'Tifblue' blueberry. 'Bluecrop' and 'Tifblue' shoots were dipped in various concentrations of kinetin solution for 5 min daily and exposed to 5 mM KNO_3. NRA was determined after 72 hr in light vs. dark. The results are means of three replications ± SE.

	NRA (nmol NO_2^-/hr/g fresh wt)			
Kinetin	Light		Dark	
Concn (µM)	Bluecrop	Tifblue	Bluecrop	Tifblue
0	9 ± 10	45 ± 4	28 ± 2	25 ± 1
1	122 ± 11	71 ± 9	32 ± 5	28 ± 3
5	193 ± 10	142 ± 14	40 ± 4	32 ± 3
10	245 ± 13	165 ± 10	58 ± 8	53 ± 4
100	90 ± 8	42 ± 2	26 ± 1	23 ± 1

been found which suggests Al^{3+} involvement in restricting the induction of nitrate transporters (Durieux et al., 1992). It has been shown that aluminum is most toxic in the presence of nitrate, and it inhibited nitrate uptake in white clover (Jarvis and Hatch, 1986), Sorghum (Galvez and Clark, 1991), rice (Van Hai et al., 1989), spruce (Peuke and Tischner, 1991), and maize (Cambraia et al., 1989).

Nitrate reductase has been shown to be stimulated by various cytokinins (Kapoor et al., 1987; Kuznetsov et al., 1979; Lu et al., 1983; Lu et al., 1990). Cytokinins have also been proven to stimulate RNA synthesis (Kuznetsov et al., 1979), affect protein synthesis (Maab and Klambt, 1979), and inhibit protein degradation (Tavares and Kende, 1970). Lu et al. (1990) demonstrated that cytokinin regulation of NR gene expression is in part controlled at the transcriptional level. Cytokinins have been found to induce NRA in tobacco (Lips and Roth-Beferano, 1969), rice (Gandhi and Naik, 1974), and barley (Lu et al., 1990). It has been reported that cytoki-

TABLE 6. Effect of various forms of cytokinins (zeatin, kinetin, 6-benzyl-aminopurine, and zeatin riboside) on 'Bluecrop' blueberry leaves. Blueberry shoots were dipped in the appropriate cytokinin solution (10 µM) twice daily for 10 min. The shoots were exposed to 5 mM KNO_3. Nitrate reductase activity was determined after 72 hr.

	NRA (nmol NO_2^-/hr/g fresh wt)	
	Light	Dark
Control	172 ± 12	65 ± 3
Zeatin	387 ± 13	133 ± 7
Kinetin	363 ± 10	105 ± 4
Zeatin riboside	293 ± 15	92 ± 5
6-benzylaminopurine	289 ± 10	88 ± 4

TABLE 7. Effect of various forms of cytokinins (zeatin, kinetin, 6-benzyl-aminopurine, and zeatin riboside) on 'Bluecrop' blueberry stems. Blueberry shoots were dipped in the appropriate cytokinin solution (10 µM) twice daily for 10 min. The shoots were exposed to 5 mM KNO_3. Nitrate reductase activity was determined after 72 hr.

	NRA (nmol NO_2^-/hr/g fresh wt)	
	Light	Dark
Control	89 ± 5	32 ± 2
Zeatin	285 ± 13	81 ± 5
Kinetin	240 ± 11	66 ± 4
Zeatin riboside	223 ± 10	60 ± 4
6-benzylaminopurine	218 ± 11	59 ± 3

nin enhancement of NRA requires nitrate and light (Lu et al., 1990). In the absence of nitrate, the induction of NRA in plants is negligible (Lu et al., 1983; Sharma and Sopory, 1987). However, kinetin has been shown to stimulate NRA in seedlings grown in the dark. When it was supplemented with nitrate, an additive effect was observed (Kapoor et al., 1987). Induction of NR with high concentrations of cytokinins yielded decreased NR activity, suggesting that high levels may be toxic to the plant.

Not only is NR a substrate-inducible and hormone-inducible enzyme, it is light-inducible as well (Hewitt, 1975). Application of cytokinins to blueberry in the dark stimulated NRA. However, in the presence of light, the level of induction with cytokinins was greater than that obtained in the dark. NRA can also be induced in cotton seedlings with exposure to light (Prakash and Kapoor, 1986). In cotton leaves, light enhanced NR activity was six times greater than in the dark (Prakash and Kapoor, 1986). Seedlings grown in the dark failed to show any significant NR activity in leaves. Light is a necessary factor for nitrate induction of NRA (Prakash and Kapoor, 1986). Apparently, light plays a crucial role in enzyme synthesis or its activation, or it may be preventing the inactivation of the enzyme. This indicates that proper functioning of the photosynthetic apparatus is necessary for complete NR induction.

Although the literature is controversial concerning the necessity for light and substrate (nitrate) when using cytokinins as an inducer for NRA, most authors agree that light, nitrate and kinetin increase NRA in an additive manner (Sharma and Sopory, 1988). This indicates that the plant uses at least two different pathways for inducing NRA. For example, when cotton seedlings were supplied with cytokinins in the dark, NRA was induced; however, in the presence of light, the level of induction was much greater. Exposing the seedlings to light increased the level of induction 2-3 fold (Prakash and Kapoor, 1986). It appears that there may be different pathways involved when plants are in the dark or exposed to light. Plant cells need a source of NADH for nitrate reduction. The source of NADH may come from glycolysis in the dark and from mitochondria in the light (Kapoor et al., 1987). Given the results obtained in this study, it appears that cytokinins may play a role in supplying enough NADH to serve as the physiological donor of electrons for *in vivo* nitrate reduction in blueberry in the dark when supplemented with nitrate. Because the light exposed plants are at times restricted to darkness, NRA is enhanced by these added cytokinins.

GROWER BENEFITS

The rate of NRA is closely associated with nitrate assimilation and ultimately to crop yield. This study indicates that 'Tifblue' is more adaptable to adverse conditions as compared to 'Bluecrop.' Light, nitrate and cytokinins stimulated NRA in both cultivars. This research provides additional information for manipulation of NRA which may enhance blueberry productivity.

LITERATURE CITED

Beck, E. and U. Renner. 1989. Ammonium triggers uptake of NO_3^- by *Chenopodium rubrum* suspension culture cells and remobilization of their vacuolar nitrate pool. Plant Cell Physiol. 30:487-495.

Cain, J.C. 1952. A comparison of ammonium and nitrate nitrogen for blueberries. Proc. Amer. Soc. Hort. Sci. 59:161-166.

Cambraia, J., J.A. Pimenta, M.M. Estevao, and R. Sontanna. 1989. Aluminum effects on nitrate uptake and reduction in sorghum. J. Plant Nutrition. 12:1435-1445.

Clarkson, D.T., L.R. Saker, and J.V. Purves. 1989. Depression of nitrate and ammonium transport in barley plants with diminished sulphate status. Evidence of co-regulation of nitrogen and sulphate intake. J. of Exp. Bot. 40:953-963.

Dirr, M.A., A.V. Barker, and D.N. Maynard. 1972. Nitrate reductase activity in leaves of highbush blueberry and other plants. J. Amer. Soc. Hort. Sci. 97:329-331.

Durieux, R.P., W.A. Jackson, E.J. Kamprath, and R.H. Moll. 1992 Inhibition of nitrate uptake by aluminum in maize. Plant Soil. 151:97-104.

Ferrari, T.E., O.C. Yoder, and P. Fhilner. 1973. Anaerobic nitrate production by plant cells and tissues: Evidence for two nitrate pools. Plant Physiol. 51:423-431.

Galvez, L. and R.B. Clark. 1991. Nitrate and ammonium uptake and solution pH changes for Al-tolerant and Al-sensitive sorghum *(Sorghum bicolor)* genotypes grown with and without aluminum. Plant Soil. 134:179-188.

Gandhi, A.P. and M.S. Naik. 1974. Role of roots, hormones and light in the synthesis of nitrate reductase and nitrite reductase in rice seedlings. FEBS Letters. 40:343-345.

Hageman, R.H. and A.J. Reed. 1980. Nitrate reductase from higher plants. Methods in Enzymology. 6:217-218.

Hewitt, E.J. 1975. Assimilatory nitrate-nitrite reduction. Annu. Rev. Plant Physiol. 26:73-100.

Holmes, R.S. 1960. Effect of phosphorous and pH on the iron chlorosis of the blueberry in water culture. Soil Science. 90:374-379.

Jarvis, S.C. and D.J. Hatch. 1986. The effects of low concentrations of aluminum on the growth and uptake of nitrate-N by white clover. Plant Soil. 95:43-55.

Kapoor, H.C., S.Prakash, and T.R. Madash. 1987. Regulation of *in vivo* nitrate reductase activity in cotton *(Gossypium hirsutum)* leaves in light and dark and the possible role of cytokinin. Indian J. Biochem. Biophys. 24:326-328.

Kuznetsov, V., V.V. Kuznetsov, and O.N. Kulaeva. 1979. Effects of cytokinin and nitrate on RNA synthesis and nitrate reductase activity in corn cockle embryo. Fiziol Rastenii. 26:309-317.

Lee, J.H. and J.S. Titus. 1992. Nitrogen accumulation and nitrate reductase activity in MM. 106 apple trees as affected by nitrate supply. J. of Hort. Sci. 67:273-281.

Lips, S.H. and N. Roth-Bejerano. 1969. Light and hormones: interchangeability in the induction of nitrate reductase. Science. 166:109-110.

Lu, J., W.Z. He, W. Chen, Y.H. Zhang, and Y.W. Tang. 1983. Studies on nitrate reductase II. Effect of benzyladenine and light on the induction of nitrate reductase in wheat seedlings. Acta Phytophysiol. Sinica. 9:41-49.

Lu, J., J.R. Ertl, and C. Chen. 1990. Cytokinin enhancement of the light induction of nitrate reductase transcript levels in etiolated barley leaves. Plant Mol. Biol. 14: 585-594.

Maab, H. and D. Klambt. 1977. Cytokinin effect on protein synthesis (in *vivo)* in higher plants. Planta. 133:117-120.

Muller, E.H. and P. Janiesch. 1993. *In vivo* nitrate reductase in *Carex pseudocyperus* L.: The influence of nitrate-ammonium concentration ratios and correlation with growth. J. Plant Nutrition. 16:1357-1372.

Oertli, E.E. 1963. Effect of form of nitrogen and pH on growth of blueberry plants. Agron. J. 55:305-307.

Pate, J.S. 1980. Transport and partitioning of nitrogenous solutes. Annu. Rev. Plant Physiol. 31:313-340.

Peuke, A.D. and R. Tischner. 1991. Nitrate uptake and reduction of aseptically cultivated spruce seedlings. *Picea abies* (L.) Karst. J. Exp. Bot. 42:723-728.

Prakash, S. and H.C. Kapoor. 1986. *In vitro* studies of nitrate reductase (NR) in cotton leaves: Role of light, carbohydrate and cytokinin in its regulation. Indian J. Biochem. Biophys. 23:143-147.

Roth-Bejerano, N. and S.R. Lips. 1970. Hormonal regulation of nitrate reductase activity in leaves. New Phytol. 69:165-169.

Sharma, A.K. and S.K. Sopory. 1987. Effect of phytochrome and kinetin on nitrite reductase activity in *Zea mays*. Plant Cell Physiol. 28:447-454.

Sharma, A.K. and S.K. Sopory. 1988. Regulation of nitrate reductase activity by phytochrome and cytokinin. Plant Physiol. Biochem. 15:107-115.

Shen, Z., Y. Liang, and K. Shen. 1993. Effect of boron on the nitrate reductase activity in oilseed rape plants. J. Plant Nutrition. 16:1229-1239.

Tavares, J. and H. Kende. 1970. The effect of 6-benzylamino-purine on protein metabolism in senescing corn leaf. Phytochemistry. 9:1763-1770.

Titus, J.S. and S.M. Hang. 1982. Nitrogen metabolism, translocation, and recycling in apple trees. Hort. Rev. 4:204-246.

Townsend, L.R. 1966. Effect of nitrate and ammonium nitrogen on the growth of lowbush blueberry. Can. J. Plant Sci. 46:209-210.

Townsend, L.R. 1967. Effect of ammonium nitrogen and nitrate nitrogen, separately and in combination, on the growth of the highbush blueberry. Can. J. Plant Sci. 47:555-562.

Van Hai, T., T. Thi Nga, and H. Laudelout. 1989. Effect of aluminum on the mineral nutrition of rice. Plant Soil. 114:173-185.

Wakhloo, J.L. 1985. Vertical profiles in concentration of potassium and growth in *Nicotiana tabacum*. Effect of potassium on metabolism proteins in successive leaves, and on the direction of translocation of ^{14}C-photosynthate and its allocation. J. Plant Physiol. 117:383-400.

Wakhloo, J.L. and A. Staudt. 1988. Development of nitrate reductase activity in expanding leaves of *Nicotiana tabacum* in relation to the concentration of nitrate and potassium. Plant Physiol. 87:258-263.

Diammonium Phosphate
Corrects Phosphorus Deficiency
in Lowbush Blueberry

John M. Smagula
Scott Dunham

SUMMARY. Eight commercial lowbush blueberry (*Vaccinium angustifolium* Ait.) fields were fertilized with diammonium phosphate (DAP) at 0, 224 or 448 kg ha^{-1} (0, 200, and 400 lbs per A) to determine the effect on subsequent leaf nutrient concentration. A RCB design with four replications was used. Leaf samples taken in July 1991 showed an increased concentration of phosphorus with increasing rate of DAP, but leaf nitrogen level was unaffected. Soil samples taken at the same time showed no significant increase in available phosphorus. Stems collected after leaf drop were taller and more branched with a larger number of flower buds due to application of DAP compared to the controls. There was a linear increase in yield with increasing rate of DAP. *[Article copies available from The Haworth Document Delivery Service: 1-800-342-9678.]*

KEYWORDS: *Vaccinium angustifolium*; Nutrition; Fertilizer

John M. Smagula is Professor of Horticulture and Scott Dunham is Crop Technician, Department of Applied Ecology and Environmental Sciences, 5722 Deering Hall, University of Maine, Orono, ME 04469-5722.

Maine Agricultural and Forest Experiment Station Contribution No. 1900.

[Haworth co-indexing entry note]: "Diammonium Phosphate Corrects Phosphorus Deficiency in Lowbush Blueberry." Smagula, John M., and Scott Dunham. Co-published simultaneously in *Journal of Small Fruit & Viticulture* (Food Products Press, an imprint of The Haworth Press, Inc.) Vol. 3, No. 4, 1995, pp. 183-191; and: *Blueberries: A Century of Research* (ed: Robert E. Gough, and Ronald F. Korcak) Food Products Press, an imprint of The Haworth Press, Inc., 1995, pp. 183-191. Single or multiple copies of this article are available from The Haworth Document Delivery Service [1-800-342-9678, 9:00 a.m. - 5:00 p.m. (EST)].

INTRODUCTION

Lowbush blueberry fields in Maine and Canada are managed native stands that are pruned biennially to maintain vigorous productive shoot growth. More fertilizer can now be used due to effective control of indigenous weed species with herbicides such as Terbacil (Trevett and Durgin, 1972; Ismail et al., 1981; Smagula and Ismail, 1981) and Hexazinone (Yarborough and Ismail, 1985; Yarborough et al., 1986; Yarborough and Bhowmik, 1989). Blueberry plants and other members of the Ericaceae family require an ammonium form of nitrogen (N) (Cain, 1952; Townsend, 1966; Townsend, 1967; Townsend and Blatt, 1966) rather than a nitrate form. The response to N fertilizer has not been consistently positive (Smagula and Ismail, 1981; Ismail et al., 1981). Previous studies reporting significant increases in yield due to added N were often conducted in fields which had no chemical weed control (Trevett, 1962). More recently, researchers have found lowbush blueberries did not respond to fertilizer (Benoit et al., 1984; Smagula and McLaughlin, 1985), perhaps due to more effective chemical weed control. As a result of removing weed competition for nutrients, lowbush blueberry plants in many fields appear to be receiving more adequate levels of nutrients provided by mineralization of soil organic matter (Smagula and McLaughlin, 1985; Smagula et al., 1987), especially N. An unpublished survey of 76 commercial lowbush blueberry fields in Maine in 1987 and 1988 indicated that 93% of the fields had leaf tissue N concentrations above the standard (1.6%) reported by Trevett (1972). The concentrations of phosphorus (P), however, were below the standard (0.125%) at 83% of the fields. Eight fields surveyed in 1987 were chosen for this experiment because they had a known history of potentially low phosphorus status. Diammonium phosphate (DAP) was chosen as a source of P because in preliminary studies P uptake was enhanced when adequate nitrogen was also available.

The objective of this study was to determine the growth and yield response of lowbush blueberries with potentially low leaf P concentrations to DAP, a fertilizer formulation containing both N and P.

MATERIALS AND METHODS

Commercial lowbush blueberry fields in eight towns: Dresden, Lamoine, Jonesboro, Calais, Columbia, Roque Bluffs, Perry and T32 MD were used in this study. In May 1991, after fields were pruned and Hexazinone was applied, 0, 224 or 448 kg ha^{-1} (0, 200, or 400 lb A^{-1}) DAP (18-46-0) was applied to experiment plots. Fertilizer treatments were replicated 4 times at each location in a randomized complete block design with plots measuring 3.05 m × 9.14 m (10 ft × 30 ft). After stems had stopped elongating, between July 1 and July 10, 1991, composite leaf samples were taken from 30 randomly cut stems within each treatment plot. Nutrient concentrations in these leaves was determined as follows: total N using the Kjeldahl method and P, K, Ca, Mg, B, Cu, Fe, Mn, Mo, Al, and Zn by plasma emission spectroscopy. Ten 7.6 cm-deep soil samples randomly sampled within each treatment plot at the time of leaf sampling in July were analyzed for organic matter (OM), pH, P, K, Mg, Ca, Zn, Mn, Fe, Al, and Cu. OM was measured by loss on ignition at 550°C for 5 hr. Soil pH was determined in distilled water at a 1:1 (w:v) ratio. Nutrients were extracted in pH 3 ammonium acetate at a 1:4 (w:v) ratio for 15 minutes and the solution analyzed by plasma emission spectroscopy.

To determine blueberry growth habits and vigor, all stems within 4 randomly selected 0.076 m^2 (0.25 ft^2) quadrates in each experimental plot were cut at ground level during October and November 1991. Stem length, branching and number of flower buds per stem were measured. Yield data were taken in August 1992 by harvesting a 0.48 m × 9.14 m (1.58 ft × 30 ft) strip in each treatment plot with a hand-held metal rake. Berries were winnowed before being weighed to obtain marketable yield.

Data were subjected to analysis of variance using the General Linear Model of SAS (SAS Institute, Inc., 1992, Cary, NC) was used.

RESULTS AND DISCUSSION

DAP application resulted in significant increases in leaf P and stem length, stem branching, flower bud formation and yield. Fertilizing with DAP did not significantly affect levels of other nutrients measured.

A linear increase (P < 0.03) in leaf P resulted from increasing rate of DAP fertilizer (Figure 1). P levels in leaf tissue of the control plots averaged 0.124% (Table 1) across all locations and varied from 0.093 to 0.145%. In two fields, control plots had leaf concentration of P above 0.130% and in these fields there was no increase in response to DAP. The average leaf N concentration in control plots (Table 1) was above the standard (1.6%) and did not increase with DAP application when data for all fields were combined. Individual fields varied in their control plot leaf N concentrations from 1.41 to 1.97%. The leaf N concentration in the field with control plots averaging 1.41% was raised to 1.70 and 1.66% by 224 and 448 kg ha^{-1}, respectively. All other nutrient elements measured (K, Ca, Mg, B, Cu, Fe, Mn, Mo, Al, and Zn) were unaffected by the fertilizer treatments (Table 1). The pH of soil samples ranged from 3.9 to 4.7 (Table 2) and was not significantly correlated with available P in control plots or plots receiving DAP. Fertilizer treatments had no effect on pH or available soil nutrients.

Stem length, number of branches and number of flower buds per stem increased linearly with increasing rate of DAP (Figure 2). That DAP caused only a small increase in branching is favorable since highly branched stems are more difficult to harvest with traditional metal hand rakes. Stem length has been positively correlated with flower bud formation and taller stronger stems may be important to hold the berries off the ground, an especially important consideration for mechanical harvesting.

Yield increased linearly with rate of DAP (Figure 3). The average increase in yield across all fields in response to applying 224 kg or 448 kg DAP/ha (200 or 400 lbs per A) was 824 kg/ha (740 lbs per A) and 1679 kg/ha (1510 lbs per A), respectively.

CONCLUSION

DAP was effective in increasing the average marketable yield across the eight fields used in this study. Since leaf P was the only nutrient influenced by DAP treatments, we conclude that the more

FIGURE 1. Leaf P concentrations as influenced by DAP fertilization. Values represent the means of eight fields (n = 32). Treatment effect was linear (P < 0.03).

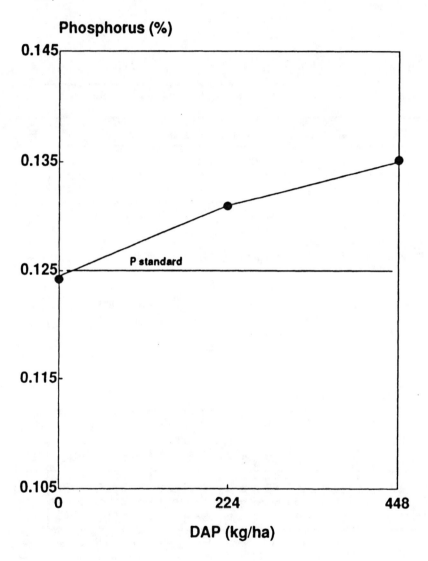

TABLE 1. Influence of DAP fertilizer on prune year leaf tissue nutrient concentrations.

Treatment

DAP

(kg ha^{-1})	(%)						(mg.kg^{-1})				
	N	K	Ca	Mg	B	Cu	Fe	Mn	Mo	Al	Zn
0	1.7^z	.47	.37	.17	20	4.6	55	994	.37	91	14
224	1.7	.48	.36	.17	21	4.4	52	1025	.33	85	16
448	1.8	.48	.39	.17	21	4.5	55	1050	.37	90	15
	NS^y	NS	NS	NS	NS	NS	NS	NS	NS	NS	NS

^{z}Mean of eight locations (N = 32).
^{y}Means within columns are nonsignificant (NS).

TABLE 2. Influence of DAP on available soil nutrients.

Treatment

DAP

(kg ha^{-1})					(mg.kg^{-1})						
	pH	OM^z	P	K	Ca	Mg	Cu	Fe	Al	Mn	Zn
0	4.4^y	15.2	16.8	62	351	54	.19	19.7	361	53	3.8
224	4.4	15.3	17.1	63	345	52	.21	19.6	383	53	3.7
448	4.3	15.3	16.0	61	360	54	.19	22.3	384	45	4.0
	NS^x	NS	NS	NS	NS	NS	NS	NS	NS	NS	NS

^{z}OM determined by loss on ignition at 550°C and expressed as percent dry weight.
^{y}Mean of eight locations (N = 32).
^{x}Means within columns are nonsignificant (NS).

FIGURE 2. Stem length, branching and flowerbud formation as influenced by Dap fertilization. Values represent the means of eight fields (n = 128). Treatment effect was linear (P < 0.0001) for all stem characteristics.

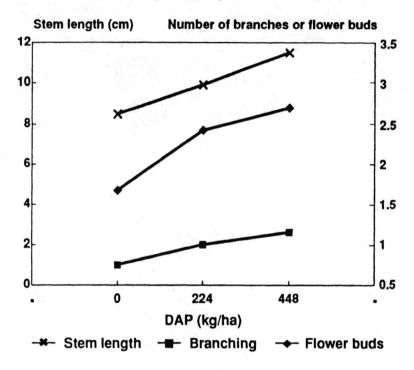

favorable P status of the plant resulted in increased plant vigor enabling the plant to grow taller and produce more flower buds.

GROWER BENEFITS

Leaf tissue analysis can guide growers in making prudent fertilizer decisions. Fertilizing with N alone when adequate soil N is available can actually result in a decrease in yield. The tradition of fertilizing every prune cycle with urea is being replaced by fertilization based on leaf tissue analysis. Growers finding leaf tissue concentrations of phosphorus below the standard (0.125%) can use DAP to supply P to effectively raise the P concentration, enhance plant vigor and increase marketable yield of lowbush blueberry.

FIGURE 3. Lowbush blueberry yield as influenced by DAP fertilization. Values represent the mean of eight fields (n = 32). Treatment effect was linear (P < 0.002).

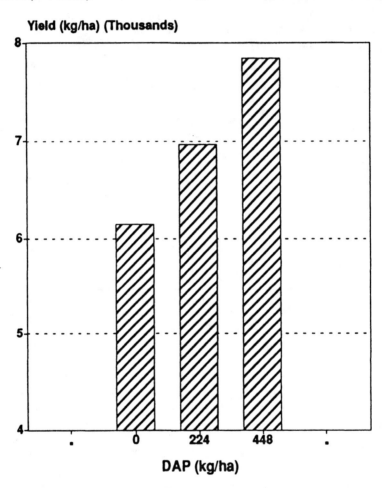

LITERATURE CITED

Benoit, G.R., W.J. Grant, A.A. Ismail, and D.E. Yarborough. 1984. Effect of soil moisture and fertilizer on the potential and actual yield of lowbush blueberries. Can. J. Plant Sci. 64:683-689.

Cain, J.C. 1952. A comparison of ammonium and nitrate nitrogen for blueberries. Proc. Amer. Soc. Hort. Sci. 59:161-166.

Ismail, A.A, J.M. Smagula, and D.E. Yarborough. 1981. Influence of pruning method, fertilizer and Terbacil on the growth and yield of the lowbush blueberry. Can. J. Plant Sci. 61:61-71.

Smagula, J.M. and A.A. Ismail. 1981. Effects of fertilizer applications, preceded by Terbacil, on growth, leaf nutrient concentration, and yield of the lowbush blueberry, *Vaccinium angustifolium* Ait. Can. J. Plant Sci. 61:961-964.

Smagula, J.M. and E.J. McLaughlin. 1985. Influence of urea fertilization on shoot development, yield, and plant spread in established lowbush blueberry fields. HortScience. 20:580.

Smagula, J.M., R. Risser, and E.J. McLaughlin. 1987. Effect of urea and alternative pruning practices on lowbush blueberry growth and yield. HortScience. 22:381.

Townsend, L.R. 1966. Effect of nitrate and ammonium nitrogen on growth of the lowbush blueberry. Can. J. Plant Sci. 46:209-210.

Townsend, L.R. 1967. Effect of ammonium nitrogen and nitrate nitrogen, separately and in combination on the growth of the highbush blueberry. Can. J. Plant Sci. 47:555-562.

Townsend, L.R. and C.R. Blatt. 1966. Lowbush blueberry: evidence for the absence of a nitrate reducing system. Plant & Soil. 25:456-460.

Trevett, M.F. 1962. Nutrition and growth of the lowbush blueberry. Maine. Agr. Expt. Sta. Bul. 605.

Trevett, M.F. 1972. A second approximation of leaf analysis standards for lowbush blueberry. Res. Life Sci. Maine Agr. Expt. Sta. 19(15):15-16.

Trevett, M.F. and R.E. Durgin 1972. Terbacil: a promising herbicide for the control of perennial grass and sedge in unplowed blueberry fields. Res. Life Sci., Maine Agric. Expt. Sta. 19:1-13.

Yarborough, D.E., J.J. Hanchar, S.P. Skinner, and A.A. Ismail. 1986. Weed response, yield and economics of Hexazinone and nitrogen use in lowbush blueberry production. Weed Sci. 34:723-729.

Yarborough, D.E. and P.C. Bhowmik. 1989. Effect of Hexazinone on weed populations and on lowbush blueberries in Maine. Fourth International Symposium on Vaccinium Culture. Acta Horticulture. 241:344-349.

RELATED POSTERS

Development of an Inoculation System for Studying Mycorrhizal Effects in Highbush Blueberry

Barbara L. Goulart
Wei Qiang Yang
Kathleen Demchak

SUMMARY. Three experiments were conducted to investigate the best methods for inoculating uninfected meristem tip-cultured 'Bluecrop' blueberry plants with an ericoid fungal symbiont (*Hymenoscyphus ericae*). Infection level was low for all treatments, and there were no statistical differences in level of infection. The control was not infected. A technique was also developed for observing the infection by *H. ericae* of highbush blueberry plantlet root cells *in vivo*. Future experiments will focus on increasing the level of infection af-

Barbara L. Goulart is Associate Professor, Wei Qiang Yang is a graduate student and Kathleen Demchak is Research Aide, Department of Horticulture, Penn State University, University Park, PA 16802.

[Haworth co-indexing entry note]: "Development of an Inoculation System for Studying Mycorrhizal Effects in Highbush Blueberry." Goulart, Barbara L., Wei Qiang Yang, and Kathleen Demchak. Co-published simultaneously in *Journal of Small Fruit & Viticulture* (Food Products Press, an imprint of The Haworth Press, Inc.) Vol. 3, No. 4, 1995, pp. 193-201; and: *Blueberries: A Century of Research* (ed: Robert E. Gough, and Ronald F. Korcak) Food Products Press, an imprint of The Haworth Press, Inc., 1995, pp. 193-201. Single or multiple copies of this article are available from The Haworth Document Delivery Service [1-800-342-9678, 9:00 a.m. - 5:00 p.m. (EST)].

193

ter inoculation. *[Article copies available from The Haworth Document Delivery Service: 1-800-342-9678.]*

KEYWORDS: Highbush blueberry; *Vaccinium corymbosum;* Nutrition; Nitrogen; Fertilization; Symbiosis; Inoculation; Techniques; Ericoid mycorrhizae

INTRODUCTION

Highbush blueberries (*V. corymbosum* L.) are currently produced on or adjacent to wetland areas, or in heavily amended soils, because of their requirement for soils with acid pH and high organic matter. While this may seem a disadvantage to growers who wish to produce highbush blueberries on typical agronomic soils, it is an adaptation that allows many of the native species of many of the *Vaccinium* to exploit relatively hostile soil environments (nutrient poor, low pH). These ericaceous plant species form ericoid mycorrhizae which, among other functions, facilitate nutrient acquisition. Mycorrhizae (which literally means "fungus roots") are the symbiotic association of plant roots with a fungus (Read, 1985). The fungus receives carbohydrates from the plant, and enhances the plant's chances for survival by aiding in nutrient uptake. The ericoid mycorrhizae are most noted for their ability to enhance nitrogen uptake; however, enhanced phosphorus uptake has also been documented (Read and Stribley, 1973; Stribley and Read, 1974). The mycorrhizal association in ericaceous plants has also been shown to mediate the effects of toxic levels of zinc and copper (Bradley et al., 1982). Further understanding of the fungal/plant interaction may enable us to exploit the relationship to expand the range of soils used for highbush blueberry production, and to better understand blueberry nutrition.

Most of the research on ericoid mycorrhizae has been conducted on seedlings of *Calluna vulgaris* (Scotch Heather) or *Vaccinium macrocarpon* (the American cranberry) (Read and Stribley, 1973; Stribley and Read, 1974; Stribley and Read, 1980; Bajwa and Read, 1986). While the fungal species which infect these plants are the same as for blueberry, a system needs to be developed for producing genetically identical infected and non-infected plants in order to study the effects of the mycorrhizae under controlled conditions. Plants obtained from commercial nurseries are invariably infected in varying degrees, and variation in infection level has been documented among blueberry cultivars (Goulart et al., 1993), so that one would also expect high levels of variation among seedling populations, making them less desirable test populations than clonal material.

The objective of the following study was to evaluate several fungal inoculation procedures for their effectiveness in producing mycorrhizal highbush blueberry plants.

MATERIALS AND METHODS

On 10 March 1994 three separate experiments were initiated to evaluate techniques for inoculating non-infected highbush blueberry plants with a known mycorrhizal fungus. For each experiment, meristem tip-cultured, rooted 'Bluecrop' plants were removed from coarse sand in which they had been rooted, the sand washed off and the roots rinsed with sterile water. Each treatment was replicated 3 times, and had 3 sub-samples, which were removed for sampling after 5, 12 or 20 days. Root samples were cleared and stained according to the technique described by Koske and Gemma (1989) and percentage infection quantified by the grid-line technique developed by Giovannetti and Mosse (1980), with the modification of examining 2 cells at 50 intersections for infection. For mycorrhizal infection level analysis, square root transformations of the original percentages were analyzed.

Experiment 1. Inoculation on petri plates. Three grooves were drilled in the sides of three petri plates each of Mitchell-Read, Malt agar, and PDA media which had colonies of *Hymenoscyphus ericae* growing on them for 51 days. The stems of the plants were wrapped with sterile cotton and placed through the grooves so the roots were on the edges of the colonies and the leaves were to the outside of the petri plates (Figure 1). A sheet of previously moistened and autoclaved filter paper was placed on the roots. Sterile water was added to each plate to saturate the filter paper and leave about 5 ml (0.15 fluid ounces) of water free. Plates were wrapped with parafilm and labeled. Three slits were made in the parafilm, each about 0.5 cm

FIGURE 1. Petri dish inoculation technique for *Vaccinium corymbosum* meristem tipcultured plantlets using the fungal symbiont *Hymenoscyphus ericae* in agar culture.

(0.2 inch) long where the top and bottom of the plate overlapped to allow water to penetrate. Each plate was then placed in a 269 × 279 mm (10.5 × 11 inches) reclosable polyethylene bag. Bottom corners of the bag were folded towards the center and taped in place to prevent the petri plate from rolling and water was added to the bag to a 1-cm (0.4 in) depth. Two holes were punched in the bag near the top center of the canopy, bags were closed and the area where the petri plates were located was covered with aluminum foil. Bags were hung from a metal frame.

Experiment 2. Inoculation in pouches. H. ericae was grown in shaken and unshaken flasks of malt liquid media for 18 days and on plates of malt agar for 51 days. In the unshaken flasks the mycelium grew in a single fluffy gray-white sheet. In the shaken flasks, the *H. ericae* formed dense gray and black aggregates. Unshaken flasks were swirled to break the mycelium into small pieces about 0.25-1.0 cm (0.1-0.4 inches) in diameter resulting in a slurry of mycelium and media. Plants were inoculated by (1) dipping the roots into the slurry (2) placing 2-3 aggregates of the mycelium from the shaken flasks among the roots or (3) placing 2-3 plugs of *H. ericae* from malt agar media among the roots. Plants were well-moistened, and pouches were sandwiched between 2 wooden garden stakes with rubber bands wrapped around the ends of the stakes to hold them together (Figure 2). Assemblies were placed across an open box which had grooves cut in the cardboard to hold them in place.

Experiment 3. Inoculation in pots. 'Bluecrop' plants were either dipped in a slurry of malt liquid media/*H. ericae* as described above, or not treated. Plants were planted in pots of coarse sand immediately.

RESULTS AND DISCUSSION

After 20 days, all treatments had relatively low levels of mycorrhizal infection (Table 1). Control plants were completely unin-

FIGURE 2. Growth pouch inoculation technique for *Vaccinium corymbosum* meristem tipcultured plantlets using the fungal symbiont *Hymenoscyphus ericae* in liquid culture.

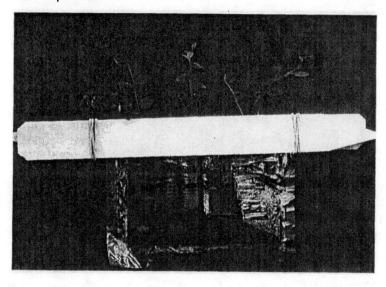

TABLE 1. Infection levels of meristem tip-cultured 'Bluecrop' highbush blueberries using several different inoculation techniques.

Treatment	% Mycorrhizal Infection After 20 Days
Growth pouches	
Mycellial dip	1.7
Mycellial solids from shaken media	2.3
Agar plugs	1.0
p(F)	0.83
Sand pots	
Mycellial dip	1.0
Control	0.0
p(F)	0.36
Petri dishes	
Potato Dextrose Agar	0.7
Malt Agar	1.5
Mitchell-Read	4.3
p(F)	0.57

fected. There were no statistical differences among treatments within each experiment, though infection levels appear to be slightly
higher in the petri dishes with Mitchell-Read media. This media
was specifically developed for mycorrhizal endophytes of *Vaccinium* and *Rhododendron* (Mitchell and Read, 1981). The blueberry
plant roots grew particularly well on all of the petri dishes (Figure 3),
and the tops of the plants looked particularly healthy on the Mitch-

FIGURE 3. Root growth of *Vaccinium corymbosum* after 20 days in petri dish inoculation culture.

ell Read media. Unfortunately, this method is the most time consuming of the three investigated.

A small growth system which consisted of two large (48 × 65 mm) cover slips which were glued 2-3 mm (0.08-0.12 inches) apart with silicon sealant was employed to watch *in vivo* infection of the blueberry plantlet roots by the fungal symbiont (Figure 4). While infection and coil development was readily observed in older cells, the young, actively growing root cells did not appear to be infected, possibly due to digestion of the hyphae by the cells (Harley, 1969). Future research will focus on increasing infection post inoculation, and modifying staining techniques to allow them to detect infection in the younger root cells.

GROWER BENEFITS

The study of the ecological success of native *Vaccinium* spp. may enable us to divert production from fragile land forms (wetlands) to

FIGURE 4. A living *Vaccinium corymbosum* root infected with *Hymenoscyphus ericae*. Magnified 1000X.

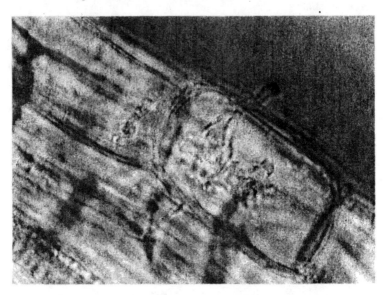

more stable systems. Additionally, we may be able to reduce fertilizer and pesticide inputs by taking advantage of the mycorrhizal relationship with respect to nutrient acquisition and pathogen interaction. In the long-term, the study of mycorrhizae as an adaptation that may be responsible for ecological success offers an alternative approach that will provide the basis for the development of many other sustainable food production systems.

LITERATURE CITED

Bajwa, R. and D.J. Read. 1986. Utilization of mineral and amino N sources by the Ericoid mycorrhizal endophyte *Hymenoscyphus ericae* and by mycorrhizal and non-mycorrhizal seedlings of *Vaccinium*. Trans. Br. Mycol. Soc. 87(2): 269-277.

Bradley, R., A.J. Burt and D.J. Read. 1982. The biology of mycorrhiza in the Ericaceae. VIII. The role of mycorrhizal infection in heavy metal resistance. New Phytol. 91:197-209.

Giovannetti, M. and B. Mosse. 1980. An evaluation of techniques for measuring vesicular arbuscular mycorrhizal infection in roots. New Phytol. 84:489-500.

Goulart, B.L., K. Demchak, M.L. Schroeder and J.R. Clark. 1993. Mycorrhizae in highbush blueberries: survey results and field response to nitrogen. HortScience 28(5):140.

Harley, J.L. 1969. The biology of mycorrhiza. Academic Press, London. pp. 171-174.

Koske, R.E. and J.N. Gemma. 1989. A modified procedure for staining roots to detect VA mycorrhizas. Mycol. Res. 92(4):486-505.

Mitchel, D.T. and D.J. Read. 1981. Utilization of inorganic and organic phosphates by the mycorrhizal endophytes of *Vaccinium macrocarpon* and *Rhododendron ponticum*. Trans. Bri. Myc. Soc. 76: 255-260.

Read, D.J. 1985. The structure and function of the vegetative mycelium of mycorrhizal roots. *In*: D.H. Jennings and A.D.M. Raynor (eds), The Ecology and Physiology of the Fungal Mycelium. Cambridge Univ. Press, Cambridge, UK pp. 215-240.

Read, D.J. and D.P. Stribley. 1973. Effect of mycorrhizal infection on nitrogen and phosphorus nutrition of ericaceous plants. Nature new biology 244:81-82.

Stribley, D.P. and D.J. Read. 1974. The biology of mycorrhiza in the Ericaceae. New Phytol. 73:1149-1155.

Stribley, D.P. and D.J. Read. 1976. The biology of mycorrhiza in the ericaceae. VI. The effects of mycorrhizal infection and concentration of ammonium nitrogen on growth of cranberry (*Vaccinium macrocarpon*, AIT.) in sand culture. New Phytol. 77:63-72.

Stribley, D.P. and D.J. Read. 1980. The Biology of mycorrhiza in the Ericacea. VII. The relationship between mycorrhizal infection and the capacity to utilize simple and complex organic nitrogen sources. New Phytol. 86:365-371.

SESSION VI:
BLUEBERRY FRUIT QUALITY
Moderator: Barbara L. Goulart

Gibberellic Acid:
A Management Tool for Increasing Yield
of Rabbiteye Blueberry

J. G. Williamson
R. L. Darnell
G. Krewer
J. Vanerwegen
S. NeSmith

J. G. Williamson and R. L. Darnell, Horticultural Sciences Department, University of Florida, Gainesville, FL 32611.

G. Krewer, Department of Horticulture, University of Georgia, P.O. Box 1209, Tifton, GA 31793.

J. Vanerwegen, Route 7, Box 1865, Ringgold, GA 30736.

S. NeSmith, Georgia Experiment Station, 1109 Experiment Street, Griffin, GA 30223-1797.

The authors thank Georgia county agents Danny Stanaland, James Clark, Mike Bruorton and John Ed Smith for assistance with the Georgia grower trials.

The authors would like to acknowledge Abbott Laboratories for their contribution of Pro-Gibb.

Mention of a proprietary product or vendor does not imply endorsement by the authors, the University of Florida, or the University of Georgia.

[Haworth co-indexing entry note]: "Gibberellic Acid: A Management Tool for Increasing Yield of Rabbiteye Blueberry." Williamson, J. G. et al. Co-published simultaneously in *Journal of Small Fruit & Viticulture* (Food Products Press, an imprint of The Haworth Press, Inc.) Vol. 3, No. 4, 1995, pp. 203-218; and: *Blueberries: A Century of Research* (ed: Robert E. Gough, and Ronald F. Korcak) Food Products Press, an imprint of The Haworth Press, Inc., 1995, pp. 203-218. Single or multiple copies of this article are available from The Haworth Document Delivery Service [1-800-342-9678, 9:00 a.m. - 5:00 p.m. (EST)].

203

SUMMARY. Field and growth chamber experiments, as well as grower trials, were conducted to determine the effects of GA_3 sprays on fruit set and yield of rabbiteye blueberry (*Vaccinium ashei* Reade) in north Florida and south Georgia. Multiple GA_3 sprays at different stages during bloom increased fruit set and yield in field experiments in Florida and in grower trials in Georgia. However, GA_3-treated fruit were smaller and matured later than untreated fruit. Similarly, GA_3 application to non-pollinated flowers under controlled environmental conditions increased percent fruit set, decreased average fruit weight, and increased fruit development period when compared to hand-pollinated fruit. High night temperature (21°C) reduced fruit set and average fruit weight compared to 10°C night temperature for both GA_3-treated and hand-pollinated fruit. Overall, results indicate that use of GA_3 under conditions of low natural fruit set in the field may significantly increase yield, although much of the yield increase would be comprised of later maturing, smaller fruit. *[Article copies available from The Haworth Document Delivery Service: 1-800-342-9678.]*

KEYWORDS: GA_3; *Vaccinium ashei*; Fruit set; Pollination

INTRODUCTION

Poor fruit set of rabbiteye blueberry is a perennial problem in the southeastern United States (Lyrene and Crocker, 1983; Mainland, 1985). The cause(s) of fruit abscission are not clear, but inadequate pollination, high temperatures during flowering, and/or insects and diseases may be contributing factors. Gibberellic acid (GA_3) has been cleared for use on blueberries to increase fruit set for over 15 years, but research results with rabbiteye blueberries have been inconclusive with respect to rate, timing and frequency of application and this growth regulator was never adopted as a commercial product for rabbiteye blueberries (Austin, 1979; Davies, 1986; Davies and Buchanan, 1979; Mainland et al., 1979). However, in 1990, field tests in south Georgia showed large increases in fruit yield from bloom applications of GA_3 to 'Climax' and 'Tifblue' rabbiteye blueberry. The varying results from the above studies suggest that timing of spray applications, or other unknown factors, may influence the effectiveness of GA_3 sprays in increasing fruit set and yield. Generally, sprays applied near full bloom were the most effective.

Several grower trials and field experiments were conducted in south Georgia and north Florida from 1990 through 1993 to determine the effect of rate and frequency of bloom and post-bloom sprays of GA₃ on rabbiteye blueberry fruit set and yield. Excessive fruit abscission in rabbiteye blueberry is often more severe in north Florida than in south Georgia. A comparison of GA₃ effects in field experiments in Florida (Davies, 1986; Davies and Buchanan, 1979) with observations of grower trials in Georgia suggest that GA₃ sprays to increase fruit set are more effective in south Georgia than in north Florida. Since air temperatures during bloom differ between the two regions, and temperatures are known to affect the extent of fruit set in a number of fruit crops (Beshir and Stevens, 1979; Calvert, 1969; Charles and Harris, 1972), a second objective of this research was to determine the effect of day/night temperature regimes on fruit set of rabbiteye blueberry receiving bloom sprays of GA₃ or hand pollination.

MATERIALS AND METHODS

1992 Florida Field Experiment. A commercial planting of six-year-old 'Beckyblue' and 'Bonita' rabbiteye blueberries, in a 1.2 m × 3.0 m spacing was used. Sprays of 250 μl · liter^{-1} GA₃ (Pro-Gibb, 4%, Abbott Laboratory, N. Chicago, IL), with 0.1% non-ionic surfactant and buffered to pH 3.1, were applied to drip with a hand gun sprayer. The first application was at 80-90% full bloom (9 March) and the second application was 10 days later. Both sprays were applied during the early evening (6:00-8:00 pm) to increase drying time. Control plants were sprayed with a water solution containing the surfactant and buffered to the same pH as the GA₃ solution. Fruit set was determined by sub-sampling shoots. Five shoots per plant were tagged, and flower number on each shoot was determined prior to spray application. Final percent fruit set was determined by counting the number of fruit on each tagged shoot on 19 May. Fruit were hand harvested over five dates, beginning 21 May and ending 1 July. At each harvest, total fruit yield (g/plant) and average fruit weight (g/fruit) were determined for each cultivar. Seasonal fruit yield and average fruit weight were determined by combining data from all harvests. The experimental design was a randomized complete block with seven replications. Five-plant plots were used (3-plant experimental units with one guard plant on either side).

1993 Florida Field Experiment. A commercial planting of 20+ year-old 'Tifblue' rabbiteye blueberry plants was used. Sprays of 250 µl · liter^{-1} GA$_3$ (Pro-Gibb, 4%), with 0.1% surfactant and buffered to pH 3.1, were applied to drip with a hand gun sprayer. There were three treatments: (1) GA$_3$ applied five days after full bloom (FB) (7 April) and again 12 days after FB (14 April); (2) GA$_3$ applied five days after FB (7 April), 12 days after FB (14 April), and 60 days after FB (8 June); and (3) a control spray (0.1% surfactant, pH 3.1) applied at all three spray dates. The 60 days after FB GA$_3$ spray was included to determine if average fruit weight could be increased by applying GA$_3$ late in the fruit development period. Fruit were hand harvested over five dates beginning 18 June and ending 22 July. At each harvest date, fruit yield (g/plant) and average fruit weight (g/fruit) were determined. Total fruit yield and average fruit weight over all harvest dates were determined. A randomized complete block design with seven replications was used. Plots contained five uniform plants within a row. Fruit were harvested from one randomly selected plant per plot, excluding the guard plants on either side of the plot.

Growth Chamber/Greenhouse Experiment. Thirty, 2-year-old, container-grown 'Beckyblue' rabbiteye blueberry plants were used. Plants were grown outdoors in a peat:pinebark (1:1 v/v) mix. In late March, 1993, flowers were thinned to approximately 60/plant. The remaining flowers were emasculated and hand pollinated with 'Climax' pollen, or emasculated and sprayed with 250 µl · liter^{-1} GA$_3$ (Pro-Gibb, 4%) with 0.1% surfactant and buffered to pH 3.1. Fifteen plants received each treatment. Immediately thereafter, plants were placed in one of three growth chambers which provided the following day/night temperature regimes: (1) 26°C (79°F) day/21°C (70°F) night; (2) 26°C (79°F) day/10°C (50°F) night; or (3) 29°C (84°F) day/10°C (50°F) night. Photoperiod was 8 h, and photosynthetic photon flux averaged 500 to 600 µmol · m^{-2} · s^{-1}. Following fruit set (4 weeks after treatment), plants were moved to a greenhouse for the duration of the experiment. Mean day/night temperatures were 26°C/23°C (79°F/73°F). Percent fruit set and average fruit weight (g/fruit) were determined. A randomized complete block design was used. Analysis of variance was used to test for temperature and GA$_3$ or pollination effects and temperature by GA$_3$ or pollination interactions. Duncan's Multiple Range test was used to separate treatment means for temperature.

Georgia Grower Trials. GA$_3$ (Pro-Gibb, 4%) sprays were applied by growers to many different cultivars under a wide array of conditions and both single and multiple applications were made during 1990 and 1991 (Table 1). All sprays were applied in the evening or at night. In all trials, rates of GA$_3$ ranged from 100 to 200 g/ha and were applied in 235 to 935 l/ha with boom or airblast sprayers. Most growers treated about 1 ha in 1990 and several treated over 20 ha in 1991.

Yield data were supplied by the growers after weighing fruit at harvest. Data from all grower trials were combined across years, cultivars, application rates, and timing, and were analyzed by analysis of variance. Linear contrasts were used for treatment comparisons.

RESULTS

1992 Florida Field Experiment. GA$_3$ increased fruit set in both rabbiteye blueberry cultivars compared to the controls (data not shown). In 'Beckyblue,' fruit set averaged 4% for the controls and 36% for the GA$_3$ treatment. Fruit set in 'Bonita' increased from 18% for the controls to 65% for the GA$_3$ treatment. Total fruit yield was also increased with the GA$_3$ sprays, but the increase was not proportional to the increase observed for fruit set. Average per bush yield over all harvests was approximately doubled for the GA$_3$ treatment compared to the control for both cultivars (Table 2). However, percent fruit set was increased 3- to 9-fold for the GA$_3$ treatments. There were many fruit which remained green and small throughout the harvest period. This was especially true for the GA$_3$-treated plants. These immature fruit were never harvested and may account for the discrepancy between fruit set increases and yield increases from the GA$_3$ treatment.

Fruit yield at each of the 5 harvest dates differed between treatments (date not shown). Fruit yield for 'Beckyblue' was the highest

TABLE 1. Effect of GA$_3$ on yield of rabbiteye blueberries in Georgia grower trials.

Year	Cultivar	Time or stage of treatment	Rate (a.i.) g/h	Yield kg/bush
1990	'Climax'	—	Untreated	2.5
		Full Bloom (FB)	100	3.9
		FB	200	5.4
		FB + 18 days After FB	100 + 100	9.7
		—	Untreated	1.2
		85% FB	100	2.1
		85% FB	100	1.8
		85% FB	200	3.7
		85% FB + 18 days After FB	100 + 100	8.1
1991	'Tifblue'	—	Untreated	1.5
		90% FB	100	3.8
		90% FB + 7 days After FB	100 + 100	4.9
	'Tifblue'			
	Field 1	—	Untreated	0.5
	Field 2	70% petal fall (PF) + 17 days after PF	100 + 100	5.4
		—	Untreated	0.6
	Field 3	70% petal fall (PF) + 17 days after PF	100 + 100	5.8
	Field 4	—	Untreated	0.5
	Field 6	—	Untreated	0.6
		70% petal fall (PF) + 17 days after PF	100 + 100	7.1
	'Woodward'			
	Field 1	—	Untreated	0.4
	Field 2	100% PF + 17 days after PF	100 + 100	2.4
		—	Untreated	0.4
		100% PF + 17 days after PF	100 + 100	2.6
	Field 3	—	Untreated	0.4
	Field 4	—	Untreated	0.4
	Field 5	—	Untreated	0.6

TABLE 2. Effect of GA_3 on seasonal fruit yield, average fruit weight and fruit development period of 'Beckyblue' and 'Bonita' rabbiteye blueberries.

	Yield (kg/bush)		Average fruit wt. (g/fruit)		FDP (days)[z]	
Treatment	'Beckyblue'	'Bonita'	'Beckyblue'	'Bonita'	'Beckyblue'	'Bonita'
Control	0.7	0.8	1.4	1.7	82	97
GA_3	1.5	1.4	1.4	1.6	96	103
Significance	*[y]	**	ns	*	*	*

[z]Fruit development period.
[y]ns, *, ** nonsignificant or significant at $P \leq 0.05$ or 0.01, respectively.

at harvest date 1 and did not differ between treatments, indicating that most of the fruit harvested at that time was pollinated. Fruit yield was greater for the GA_3 treatment than for the control at harvest dates 2 through 5. For 'Bonita,' the major fruit harvest occurred at harvest date 4 for both treatments. As with 'Beckyblue,' there were no differences in yield at harvest date 1, but the GA_3 treatment resulted in significant increases in yield compared to the control at harvest dates 2 through 4.

Average fruit fresh weight for 'Beckyblue,' averaged for all harvest dates, was not affected by GA_3, but 'Bonita' fruit weight was slightly less for the GA_3 treatment than for the control (Table 2). Average fruit weight of 'Bonita' was reduced for the GA_3 treatment when compared to the control at harvest dates 3 through 5, contributing to the overall (seasonal) decrease in fruit weight observed for 'Bonita' (data for individual harvest dates not shown).

1993 Florida Field Experiment. Two or three sprays of GA_3 increased seasonal fruit yield of 'Tifblue' rabbiteye blueberry when compared to the control (Table 3). The relative increase in fruit yield per plant for 'Tifblue' (ca 2.5x) was similar to that observed for 'Beckyblue' and 'Bonita' during 1992.

Both GA_3 treatments resulted in higher fruit yields than the control at all but the first harvest date (Figure 1). The greatest increase in fruit yield from the GA_3 treatments occurred during harvest dates

TABLE 3. Effect of GA₃ on seasonal fruit yield and average fruit weight of 'Tifblue' rabbiteye blueberry.

Treatment	Yield (kg/bush)	Average fruit weight (g/berry)	FDP[z]
Control	3.7 b[y]	1.3 a	93 b
GA₃[x](2 app.)	9.7 a	0.9 b	98 a
GA₃[w](3 app.)	8.3 a	0.9 b	98 a

[z]Fruit development period.
[y]Mean separation by Duncan's Multiple Range test, 0.05%.
[x]GA₃ ($\mu l \cdot liter^{-1}$) sprayed to drip 5 and 12 days after full bloom.
[w]GA₃ ($\mu l \cdot liter^{-1}$) sprayed to drip 5, 12 and 60 days after full bloom.

FIGURE 1. Effect of GA₃ sprays on fruit yield per plant, by harvest date, of 'Tifblue' rabbiteye blueberry. Mean separation within harvest date by Duncan's Multiple Range test, $P \leq 0.05$.

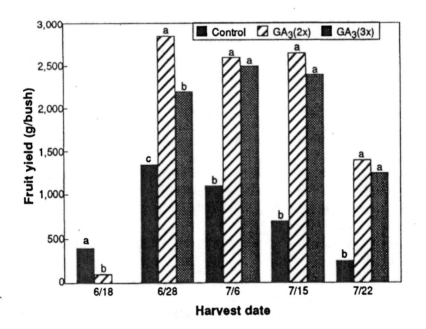

4 and 5. Approximately three to six times as much fruit was harvested from the GA3-treated plants as from the control plants during this period. Total fruit yield, and fruit yield at most harvest dates, did not differ between the two GA3 treatments.

As in 1992, average fruit fresh weight decreased throughout the harvest season regardless of treatment (Figure 2). Average fruit weight for each harvest date was less for both of the GA3 treatments compared to the control. The seasonal average fruit fresh weight was significantly less for the GA3-treated fruit compared to the control (Table 3). The additional GA3 spray, applied 60 days after bloom, did not increase average fruit weight when compared to the more conventional bloom plus post-bloom split-application of GA3.

Growth Chamber/Greenhouse Experiment. No temperature by pollination treatment interactions were found. The GA3 treatment

FIGURE 2. Effect of GA3 sprays on average fruit fresh weight, by harvest date, of 'Tifblue' rabbiteye blueberry. Mean separation, within harvest dates, by Duncan's Multiple Range test, P ≤ 0.05.

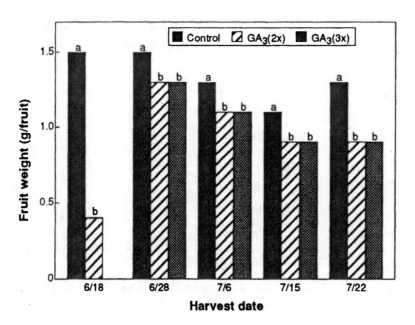

increased percent fruit set compared to hand pollination (Table 4). However, average fruit weight was less for GA_3-treated plants than for hand-pollinated plants even with low fruit loads of less than 60 berries per plant. Average fruit development period (FDP) was increased by 10 days for the GA_3 treatment compared to hand pollination. This GA_3-induced delay in fruit maturity is consistent with field observations.

Day/night temperatures affected percent fruit set and average fruit fresh weight. The 26°C/10°C (79°F/50°F) temperature regime increased fruit set relative to the 26°C/21°C (79°F/70°F) treatment and increased average fruit weight compared to either of the other two temperature regimes. Thus, percent fruit set was reduced by high night temperature (21°C [70°F]), and average fruit weight was reduced by high night (21°C [70°F]) or high day (29°C [84°F]) temperatures.

Georgia Grower Trials. Single applications of GA_3 at 100 or 200 g/ha and split applications of GA_3 at 100 g/ha significantly in-

TABLE 4. Temperature and pollination effects on fruit set and development of 'Beckyblue' rabbiteye blueberry.

Main effect	Fruit set (%)^z	Average fruit wt. (g)	FDP^y (days)
Day/night temperature (°C)			
26/21	63.9 b^x	1.5 b	85 b
26/10	83.2 a	1.7 a	88 ab
29/10	71.4 ab	1.4 b	90 a
GA_3/Pollination			
GA_3	79.8 a	1.2 b	93 a
Pollination	66.1 b	1.9 a	83 b

^{z}Data were arc sin transformed prior to analysis.
^{y}FDP = fruit development period.
^{x}Mean separation by Duncan's Multiple Range test, 5% level (temperature) and t-test, 5% level (pollination).

creased fruit yield compared to the control (Table 5). Yields were usually increased 4 to 7 fold, during years with poor fruit set (1990 and 1991). In a year with good natural fruit set (1992), yields were increased 15-25% based on estimates of the number of seedless or nearly seedless berries per bush (data not shown).

DISCUSSION

Results from previous studies on the usefulness of GA_3 sprays to increase fruit set and yield of rabbiteye blueberry are contradictory (Mainland et al., 1979; Davies and Buchanan, 1979; Davies, 1986). Conflicting results may be due, in part, to differences in rate, timing, and frequency of application. Recently, NeSmith and Krewer (1992) identified stage 5 of rabbiteye flower bud phenology (i.e.,

TABLE 5. Effect of GA_3 on seasonal fruit yield of rabbiteye blueberry cultivars in Georgia grower trials.

Treatment	Fruit yield (kg/bush)
1. Control	0.8
2. GA_3 – 100 g/h	2.9
3. GA_3 – 200 g/h	4.5
4. GA_3 – 100 g/h + 100 g/h	5.7
Contrasts	
1 vs. 2	**
1 vs. 3	**
1 vs. 4	**
2 vs. 3	ns
2 vs. 4	+
3 vs. 4	ns

+, ** significant at P ≤ 0.10 and 0.01, respectively.

just prior to corolla opening [Spiers, 1978]) as the optimum stage for GA_3 sprays to increase fruit set. In our studies, both fruit set and yield of rabbiteye blueberries were increased in commercial plantings by single or multiple GA_3 sprays, timed to begin when the majority of flowers were at, or close to, stage 5 of flower bud development. However, rabbiteye cultivars may respond differently to GA_3 sprays. The two rabbiteye cultivars included in the 1992 Florida field experiment often set light crops and GA_3 sprays did not completely alleviate this problem. Although yield of GA_3-treated bushes increased about 100% over the control, yields were relatively low for all treatments and were far below the maximum carrying capacity of the bushes. When GA_3 was applied to 'Tifblue' in north Florida, yield increased about 150% over the control. Yields of the controls were relatively high in the 1993 Florida experiment (about 4000 g/plant [8.8 lbs]) indicating that natural fruit set was high. In that experiment, GA_3-treated plants averaged about 9000 g/plant (19.8 lbs), which may have been approaching the maximum carrying capacity of the bushes. In Georgia grower trails, fruit yields were increased much more by GA_3 sprays (relative to non-treated plants) when natural fruit set was low than when natural fruit set was high. In our experiments, the effects of GA_3 bloom sprays to increase fruit set and yield of rabbiteye blueberry varied with year and location. However, results were more promising than those reported previously with rabbiteye blueberries (Austin, 1979; Davies, 1986; Davies and Buchanan, 1979). Overall, the response to bloom sprays of GA_3 appeared to be similar for north Florida and south Georgia.

GA_3 sprays decreased average fruit fresh weight in field and greenhouse studies. A significant (0.3 g/fruit) decrease in fruit weight was observed in the greenhouse study, even when flowers were thinned to approximately 60 per plant. A similar decline in average fruit weight (0.4 g/fruit) for GA_3-treated fruit vs. control fruit was observed for 'Tifblue' where flowers and fruit were not thinned and yields of GA_3-treated plants were high. This suggests that reduced fruit weight observed with GA_3-treated fruit is a direct effect of the treatment and not a response to heavier crop loads. The third application of GA_3, 60 days after bloom, did not increase fruit weight compared to the control or split-application treatments. In contrast, post fruit-set applications of GA$_3$ have been reported to increase fruit weight of seedless grapes in Florida (Halbrooks and Mortensen, 1988). However, application timing is critical in eliciting growth regulator responses, and additional studies on timing of post fruit-set GA$_3$ sprays to increase fruit size in blueberry are warranted.

FDP was increased for GA$_3$-treated fruit in the 1992 and 1993 Florida field experiments and in the greenhouse experiment. The FDP for GA$_3$-treated fruit was from 6 ('Bonita') to 14 ('Becky-blue') days longer in 1992, and 5 days longer ('Tifblue') in 1993 than for the control fruit. In the greenhouse study, FDP was increased by 10 days for GA$_3$-treated 'Beckyblue' fruit over hand-pollinated fruit. This agrees with work by Krewer et al. (1991) and with observations made throughout north Florida and south Georgia where GA$_3$ has been used in commercial plantings.

Air temperature during pollination, or during GA$_3$ application, may play an important role in fruit set. Both fruit set and average fruit weight were greater for the 26°C (79°F) day/10°C (50°F) night temperature regime than for the 26°C (79°F) day/21°C (70°F) night temperatures. Similar deleterious effects of high temperatures on fruit set have been reported for tomato (Beshir and Stevens, 1979; Calvert, 1969; Charles and Harris, 1972). However, exogenous applications of benzylamino-purine (BA) or BA + GA$_{4+7}$ increased fruit set of tomato under high temperature conditions. In our experiment, application of GA$_3$ did not overcome the adverse effects of high night temperature on fruit set of rabbiteye blueberry. The decrease in fruit set and fruit size from 21°C (70°F) night temperature vs. 10°C (50°F) night temperature were not exceptionally large, but taken together, they could result in significant decreases in yield. The detrimental effect of high temperatures on blueberry fruit set and development may partially explain the differences in yield of rabbiteye blueberries in north Florida versus south Georgia.

Two applications of GA$_3$, beginning at stage 5 of flower bud development, increased fruit set, yield and FDP of several rabbiteye blueberry cultivars. The increases in yield of GA$_3$-treated plants relative to controls were similar for south Georgia and north Florida, but they varied considerably with location and year. Presumably, the amount of natural fruit set influenced the effectiveness of GA$_3$ sprays at increasing yields over the controls. GA$_3$-treated fruit had fewer seed, and were smaller and delayed in maturity compared to non-treated fruit. The longer FDP of GA$_3$-treated fruit may be more critical for north Florida than for south Georgia since Florida growers pack and ship primarily for the early fresh market.

GROWER BENEFITS

Current recommendations call for initial GA$_3$ application when the highest percentage of blooms are at stage 5 of bloom phenology (Spiers, 1978). A cultivar directed treatment (CDT) of application (Table 6) using an airblast sprayer with treatments spaced about seven days apart has produced the best results in Georgia grower trials (Table 1). This is probably due to more flowers receiving one or two applications of GA$_3$ at a receptive stage and the benefits of the second and third sprays in helping hold the fruit initially "set."

GA$_3$ applications, especially where more than one spray was applied, generally decreased fruit weight, delayed ripening, and decreased soluble solids slightly, but overall quality was enhanced, especially in years of poor natural fruit set where fruit were predominately seedless, or nearly seedless (Krewer et al., 1991; Vanerwegen and Krewer, 1990).

GA$_3$ has inhibited flower bud formation in many crops including rabbiteye blueberries (Mainland et al., 1979; Vanerwegen and Krewer, 1990). However, return bloom has not been a problem under Georgia or Florida conditions except in cases of overcropping bushes low in vigor. Current suggestions for commercial use of GA$_3$ on rabbiteye blueberries in Georgia and Florida are outlined in Table 6.

TABLE 6. Suggestions for the use of gibberellic acid on rabbiteye blueberries.

Response	Material	Timing	Rate of Material	Remarks
Increase Fruit Set	Gibberellic acid- ProGibb 4% liquid concentrate or GibGro 4 LS	First application: When the highest percentage of blooms are elongated, but not yet open (stage 5). At this stage some flowers will be open. Second application: 10-14 days later	200 g GA$_3$/ha (80 g GA$_3$/a)	1. Minimum 40 gal. water/a. 2. Add 1 pt. of non-ionic surfactant/a. 3. If solution is alkaline (pH greater than 8.0), lower the pH with a buffering agent.

Additional considerations and precautions for using gibberellic acid.

1. Good results can be obtained with alternate row middle (ARM). Where two varieties with different bloom dates are planted together this may be the best method. It requires an air-blast type sprayer.

   Using CDT, the first and second application of GA$_3$ are directed toward the first cultivar to bloom. Some GA$_3$ will also reach the adjacent cultivar, helping the early flowers to set. The third and fourth sprays are directed towards the later blooming cultivar. The last flowers/fruit to open will benefit from spray drift from this late blooming cultivar. Applications using this method are usually applied seven to 10 days apart. The total amount of GA$_3$ sprayed each time is 50 g/ha (20 g/a) since the sprayer travels only alternate rows or the equivalent each time. The total applied during the season is 200 g/ha (80 g/a).

2. Night applications are one way to increase drying time. Blueberries are waxy and a slow drying time may assist penetration.

3. GA$_3$ should be compatible with most fungicides, but a small-scale trial is recommended to make sure settling or is not a problem.

4. Do not apply within 40 days of harvest.

5. If possible, do not apply if rain is forecast within 18 hours.

6. Do not apply to bushes in a low state of vigor.

7. Excessive (200 g/ha) or late applications of GA$_3$ may reduce flowering the following year. This is especially true if the bushes are in a low state of vigor.

LITERATURE CITED

Ahmadi, A.B.E. and M.A. Stevens. 1979. Reproductive responses of heat-tolerant tomatoes to high temperatures. J. Amer. Soc. Hort. Sci. 104(5) 686-691.

Austin, M.E. 1979. The effect of gibberellic acid on rabbiteye blueberries. Georgia Agr. Res. 20:8-10.

Calvert, A. 1969. Studies on post-initiation development of flower buds of tomato (*Lycopersicon esculentum*). J. Hort. Sci. 44:117-126.

Charles, W.B. and R.E. Harris. 1972. Tomato fruit set at high and low temperature. Canadian J. Plant Sci. 52:497-506.

Davies, F.S. 1986. Flower position, growth regulators, and fruit set of rabbiteye blueberries. J. Amer. Soc. Hort. Sci. 111:338-341.

Davies, F.S. and D.W. Buchanan. 1979. Influence of GA_3 on rabbiteye blueberry fruit set, yield, and quality, p. 229-236. In: J.N. Moore (ed). Proc. IV North American Blueberry Res. Workers Conf., Fayetteville, Ark.

Halbrooks, M.C. and J.A. Mortensen. 1988. Effects of gibberellic acid on berry weight and seed development in 'Orlando' seedless grape. HortScience 23:409.

Krewer, G., D.S. NeSmith, D. Stanaland, J. Clark, M. Bruorton, and J.E. Smith. 1991. Results of the 1991 field trial with gibberellic acid on rabbiteye blueberries in south Georgia. Proc. S.E. Prof. Fruit Workers Conf. p. 9.

Lyrene, P.M. and T.E. Crocker. 1985. Poor fruit set on rabbiteye blueberries after mild winters; possible causes and remedies. Proc. Fla. State Hort. Soc. 96:195-197.

Mainland, C.M. 1985. Some problems with blueberry leafing, flowering, and fruiting in a warm climate. Acta Hort. 165:29-34.

Mainland, C.M., J.T. Ambrose, and L.E. Garcia. 1979. Fruit set and development of rabbiteye blueberries in response to pollinator cultivar or gibberellic acid, p. 203-211. In: J.N. Moore (ed.). Proc. IV North American Blueberry Res. Workers Conf., Fayetteville, Ark.

NeSmith, D.S. and G. Krewer. 1992. Flower bud stage and chill hours influence the activity of GA_3 applied to rabbiteye blueberry. HortScience 27:316-318.

Spiers, J.M. 1978. Effect of stage of bud development on cold injury in rabbiteye blueberry. J. Amer. Soc. Hort. Sci. 103:452-455.

Vanerwegen, J. and K.G. Krewer. 1990. Gibberellic acid has potential for use in rabbiteye blueberries. Proc. of S.E. Prof. Fruit Workers Conf. p. 5.

Diseases of Blueberry Fruit at Harvest in North Carolina

W. O. Cline
R. D. Milholland

SUMMARY. Blueberries were harvested from 11 cultivars and four breeding selections from four locations in 1989 and 1990. Annual disease losses at harvest averaged 9.6% and were primarily due to five diseases: Mummy berry (*Monilinia vaccinii-corymbosi*) 5.6%, phomopsis soft rot (*Phomopsis vaccinii*) 2.9%, phyllosticta rot (*Phyllosticta vaccinii*) 0.4%, ripe rot (*Colletotrichum* sp.) 0.4% and alternaria rot (*Alternaria tenuissima*) 0.2%. Phomopsis soft rot occurred both as a localized calyx-end rot and as a soft rot detectable only by feel. Phyllosticta rot is an early season disease, and 2/3 of the infected fruit were collected at the first harvest date in 1990. Significant differences in disease levels occurred among cultivars and locations. Low levels of ripe rot and alternaria rot were attributed to a 7-day harvesting interval. A previously unreported disorder in the cultivar Cape Fear resulted in soft, unmarketable fruit. *[Article copies available from The Haworth Document Delivery Service: 1-800-342-9678.]*

KEYWORDS: Mummy berry; *Monilinia vaccinii-corymbosi*; Ripe rot; *Colletotrichum*; *Phomopsis*; *Alternaria*

W. O. Cline is Researcher/Extension Specialist and R. D. Milholland is Professor, Department of Plant Pathology, Box 7616, North Carolina State University, Raleigh, NC 27695.

[Haworth co-indexing entry note]: "Diseases of Blueberry Fruit at Harvest in North Carolina." Cline, W. O., and R. D. Milholland. Co-published simultaneously in *Journal of Small Fruit & Viticulture* (Food Products Press, an imprint of The Haworth Press, Inc.) Vol. 3, No. 4, 1995, pp. 219-225; and: *Blueberries: A Century of Research* (ed: Robert E. Gough, and Ronald F. Korcak) Food Products Press, an imprint of The Haworth Press, Inc., 1995, pp. 219-225. Single or multiple copies of this article are available from The Haworth Document Delivery Service [1-800-342-9678, 9:00 a.m. - 5:00 p.m. (EST)].

INTRODUCTION

A number of fungi are known to cause fruit rotting diseases of blueberries (*Vaccinium corymbosum* L., *V. ashei* Reade), either before harvest or in post-harvest marketing channels (Cappellini et al. 1982; Eck 1988; Milholland 1984; Milholland and Daykin 1983). These disease problems are most often cultivar-specific (Cline 1990; Eck 1988; Ramsdell 1983) and can be affected by climate, fruit maturity at harvest, and post-harvest temperature (Eck 1988; Hruschka and Kushman 1963; Milholland and Jones 1972). Rots caused by *Colletotrichum, Botrytis, Phomopsis* and *Alternaria* sp. fungi are cited as primary causes of post-harvest problems (Cappellini et al. 1982); the proportion of rot caused by each fungus appears to vary from one blueberry-producing state to the next.

Several local examples of the effects of cultivar, season, and location are evident in southeastern North Carolina. The cultivar Harrison is no longer recommended for planting due to susceptibility to phomopsis fruit rot caused by *Phomopsis vaccinii* Shear. (Cline 1990; Milholland and Daykin 1983). Mummy berry (*Monilinia vaccinii-corymbosi* [Reade] Honey) and ripe rot (*Colletotrichum* sp.) vary with changes in seasonal or weekly rainfall, respectively; mummy berry requires adequate soil moisture for apothecium development (Milholland 1974), while ripe rot spores require free water on the berry surface in order to germinate and infect (Eck 1988).

Location effects can be seen (and caused) on a national scale by differences in cultivar recommendations (Ballington and Krewer 1989; Doughty et al. 1988; Milholland 1984; Ramsdell 1983). In North Carolina, severity of disease varies among fields affected by stem blight (*Botryosphaeria dothidea* (Moug.: Fr.) Ces. & de Not.) (Creswell 1987). Farm-to-farm variation in the type and severity of fruit diseases may also be significant.

This study was initiated in 1989 to determine the type and prevalence of fruit rot diseases occurring on blueberry varieties grown in southeastern North Carolina, and to determine the fruit rot susceptibility status of new cultivars (Blue Ridge, Bounty, Cape Fear, O'Neal and Reveille) and one advanced selection (NC 2161).

MATERIALS AND METHODS

Four locations were chosen in southeastern North Carolina for inclusion in the survey based on the presence of the cultivars needed to complete the survey. These locations were: (1) White Lake in Bladen County, (2) Ideal Tract at the Horticultural Crops Research Station in Castle Hayne, New Hanover County, (3) New Bern-Bridgeton area in Craven County, and (4) Charity in Duplin County. *Vaccinium corymbosum* L. cultivars surveyed included highbush ('Bluechip,' 'Bounty,' 'Croatan,' 'Harrison,' 'Murphy' and 'New Murphy') and southern highbush ('Blue Ridge,' 'Cape Fear,' 'O'Neal,' 'Reveille' and NC 2161). Rabbiteye (*V. ashei* Reade) cultivars surveyed were 'Premier,' 'Powderblue' and 'Tifblue.' All plots were harvested during wet (1989) and dry (1990) seasons.

Three-bush plots were selected for each cultivar × location prior to the harvest season in 1989; the same plots were used in 1990. Harvest was accomplished by picking all ripe fruit from plots at each harvest date, with two pickings 1 wk apart on each plot. Fruit was not allowed to reach an over-ripe stage before harvest. Timing of the first harvest was based on the availability of a minimum of 1 L (1 qt) of fruit from each bush. Bladen county plots were picked first due to slightly earlier ripening followed by New Hanover, Duplin and Craven counties on successive days. Sampling began on 31 May in 1989 and on 15 May in 1990. Rabbiteye cultivars were harvested beginning on 7 July in 1989 and on 13 June in 1990. Fruit from each of the bushes in a given plot was bulked together in the field and transported to the lab for overnight storage at 1°C (34°F). The following day each bulk sample was individually hand-sorted to separate out diseased fruit. Diseased fruit were separated again into categories: Mummy berry, ripe rot, phomopsis rot, alternaria rot, phyllosticta rot, and unknown. Berries in the unknown category were plated on acidified Potato-Dextrose Agar (aPDA) and evaluated for the presence of pathogenic fungi after 7 days under continuous fluorescent lighting at 20°C (68°F). In 1989, no replication of location × cultivar samples was attempted. In 1990, four equal samples were harvested from each plot and evaluated separately.

RESULTS AND DISCUSSION

Mummy berry caused by *Monilinia vaccinii-corymbosi* was clearly the greatest yield-reducing disease factor at harvest, followed by phomopsis and phyllosticta fruit rots (Table 1). These three diseases accounted for 93% of all diseased berries at harvest. Ripe rot and alternaria rot, both serious post-harvest decay organisms, caused very little disease by harvest time when bushes were picked clean using a 7-day picking interval.

Susceptibility of cultivars to specific diseases is listed in Table 2. Of particular interest are the recently released 'Bounty,' 'Cape Fear,' 'O'Neal' and 'Reveille.' 'Bounty' appears to be extremely susceptible to mummy berry. 'Cape Fear' was found to have an unexplained soft fruit disorder that cannot be attributed to a disease; this will require further investigation before this cultivar can be recommended for planting. Variation in disease incidence by location was most significant for mummy berry. Differences in fungicide programs, soil types and soil moisture contributed to this variation (Table 3).

TABLE 1. Survey results of highbush blueberry yield losses for wet (1989) and dry (1990) harvest seasons.

Pathogen	Common Name	Loss per year (%)		
		1989	1990	Average
Monilinia vaccinii-corymbosi	Mummy berry	7.2	4.1	5.6
Colletotrichum sp.	Ripe rot	0.4	0.4	0.4
Phomopsis vaccinii	Phomopsis fruit rot	2.8	3.0	2.9
Alternaria tenuissima	Alternaria rot	0.1	0.3	0.2
Phyllosticta vaccinii	Phyllosticta fruit rot	0.5	0.4	0.4
Other[z]		<0.1	<0.1	<0.1
Total		11.0	8.2	9.6

[z]Including *Botryosphaeria, Pestalotia* and *Gloeosporium* sp.

TABLE 2. Occurence of pathogens found in North Carolina fields on nine highbush blueberry cultivars and one advanced selection, 1989-90.

	Average Disease Loss (%)				
Cultivar	Monilinia	Colletotrichum	Phomopsis	Alternaria	Phyllosticta
'Bluechip'	5.4	0.2	1.6	0.3	0.4
'Croatan'	5.4	0.0	2.8	0.3	0.3
'Harrison'	9.1	0.3	4.6	0.4	0.8
'Murphy'	6.4	0.8	3.0	0.0	0.6
'Cape Fear'[z]	1.3	0.8	3.0	0.2	0.4
'Bounty'	17.2	0.2	1.4	0.2	0.4
'O'Neal'	1.4	0.1	2.6	0.1	0.5
'New Murphy'	1.1	0.7	3.1	0.0	0.8
'NC 2161	1.3	0.2	1.0	0.2	0.0
'Reveille'	1.4	0.2	2.2	0.2	0.2
Average[y]	5.0	0.4	2.5	0.2	0.4

[z]'Cape Fear' exhibits an unexplained softness radiating from the stem scar which damaged 5.4% of the fruit of this variety in addition to the losses listed as due to disease. To date, no pathogen has been isolated from the affected tissue.
[y]Total averages vary from Table 1 due to the lack of individual cultivar data from some locations.

CONCLUSION

Primary post-harvest decay organisms (*Alternaria, Colletotrichum*) are not responsible for significant yield reduction prior to harvest in North Carolina when growers are able to harvest every 7 days. Cultivars surveyed did not appear to be very susceptible to either organism. Of the diseases causing > 1% loss at harvest, mummy berry was the most prevalent, but also appeared to be significantly reduced through the use of triforine sprays.

TABLE 3. Disease incidence by location for 1989-90 harvest seasons.

County	Average Disease Loss (%)				
	Mummy berry[z]	Ripe rot	Phomopsis[y]	Alternaria	Phyllosticta
Bladen	7.5	0.4	3.9	0.3	0.2
New Hanover	3.9	0.4	2.5	0.0	0.4
Craven	7.9	0.2	2.6	0.1	0.8
Duplin	0.2	0.2	1.6	0.2	0.3
Average[x]	4.9	0.3	2.6	0.2	0.4

[z]Bladen and Craven locations received no triforine sprays for mummy berry control. The New Hanover location was sprayed twice each year by aerial applications prior to bloom. The Duplin location received ground applications of triforine, and has much less organic content in the soil, which may account for the low incidence of mummy berry.
[y]Post-bloom sprays (captafol, folpet or captan) were applied to all locations except Bladen in both years.
[x]Total averages for each disease vary from table to table due to the absence of some cultivars at some locations.

GROWER BENEFITS

Blueberry producers in North Carolina who train pickers to exclude mummied berries and *Phomopsis*-affected (soft) fruit have the ability to pack fruit with very low levels of disease or decay. However, growers must be able to maintain the all-important 7-day (or less) interval between harvests in a given field, and pickers must thoroughly harvest all ripe fruit before moving on to the next bush. New North Carolina cultivars have been evaluated and disease susceptibilities identified. The importance of triforine use for mummy berry control and the prevalence of this disease in four locations has been determined.

LITERATURE CITED

Ballington, J. R. and Krewer, G. W. 1989. Blueberry Culture. *In* Small Fruit Pest Management and Culture. University of Georgia Cooperative Extension Service Bulletin 1022. pp. 16-28.
Cappellini, R. A., Ceponis, M. J. and Koslow, G. 1982. Nature and Extent of

Losses in Consumer-grade Samples of Blueberries in Greater New York. HortScience 17:55-56.

Cline, W. O. 1990. Blueberry Disease Control in 1990–With Limited Supplies of Difolatan. *In* Proceedings 24th Annual Open House, Southeastern Blueberry Council, Clinton, NC. January 25, 1990. pp. 17-19.

Creswell, T. C. 1987. Occurrence and Development of Stem Blight of Blueberry in North Carolina Caused by *Botryosphaeria dothidea*. Ph.D. thesis. North Carolina State University, Raleigh.

Doughty, C. C., Adams, E. B. and Martin, L. W. 1988. Highbush Blueberry Production in Washington and Oregon. Pacific North West Cooperative Extension Bulletin No. 215. 25 pp.

Eck, Paul. 1988. Disease Control. *In* Blueberry Science, Rutgers University Press, New Brunswick, NJ. pp. 180-200.

Hruschka, H. W. and Kushman, L. J. 1963. Storage and Shelf Life of Packaged Blueberries. U. S. Department of Agriculture Marketing Research Report No. 612. 15 pp.

Milholland, R. D. 1984. Blueberry Diseases. *In* Diseases and Arthropod Pests of Blueberries. North Carolina Agricultural Research Service Bulletin 468. pp. 1-11.

Milholland, R. D. 1974. Factors Affecting Apothecium Development of *Monilinia vaccinii-corymbosi* From Mummied Highbush Blueberry Fruit. Phytopathology 64:296-300.

Milholland, R. D. and Daykin, M. E. 1983. Blueberry Fruit Rot Caused by *Phomopsis vaccinii*. Plant Disease 67:325-326.

Milholland, R. D. and Jones, R. K. 1972. Postharvest Decay of Highbush Blueberry Fruit in North Carolina. Plant Disease Reptr. 56:118-122.

Ramsdell, D. C. 1983. Blueberry Diseases in Michigan. Michigan State University Cooperative Extension Service Bulletin E-1731. pp. 1-7.

Blueberry Culture and Research in Japan

T. Tamada

SUMMARY. In Japan, 16 species of *Vaccinium* plants grow wild that have not been improved through breeding for commercial growing. The introduction of blueberries into Japan began in 1951; however, commercial blueberry culture made slow progress. In 1991, highbush and rabbiteye blueberries were planted on about 190 hectare (469 A).

One of the severe problems of blueberry culture in Japan is the ripening of the main highbush blueberry cultivars during the rainy season (June and July). For this reason, it is a very difficult to produce fruit of high quality.

The primary areas of blueberry research in Japan are as follows: (1) variety tests of highbush and rabbiteye blueberries, (2) techniques and appropriate methods of training and pruning, (3) mineral nutrition and diagnosis of nutrient condition, (4) short-term storage tests and keeping quality in the rainy season and mid-summer, and (5) protected cultivation of highbush blueberries. *[Article copies available from The Haworth Document Delivery Service: 1-800-342-9678.]*

KEYWORDS: Japanese blueberry culture; Research

WILD SPECIES OF VACCINIUM IN JAPAN

There are 16 species of *Vaccinium* that grow wild in Japan (Makino, 1970; Ohwi and Kitagawa, 1983; Uehara, 1990) (Table 1).

T. Tamada is Fruit Scientist and Sub-Chief of the Fourth-year Course, Chiba-ken Agricultural College, 1059 Ienoko, Togane-shi, Chiba-ken 283, Japan.

[Haworth co-indexing entry note]: "Blueberry Culture and Research in Japan." Tamada, T. Co-published simultaneously in *Journal of Small Fruit & Viticulture* (Food Products Press, an imprint of The Haworth Press, Inc.) Vol. 3, No. 4, 1995, pp. 227-241; and: *Blueberries: A Century of Research* (ed: Robert E. Gough, and Ronald F. Korcak) Food Products Press, an imprint of The Haworth Press, Inc., 1995, pp. 227-241. Single or multiple copies of this article are available from The Haworth Document Delivery Service [1-800-342-9678, 9:00 a.m. - 5:00 p.m. (EST)].

TABLE 1. Wild species of *Vaccinium* in Japan.*

Vaccinium species	Japanese common name
V. axillare Nakai	Kurousugo
V. bracteatum Thunb.	Shashanbo
V. boninense Nakai	Muninshashanbo
V. ciliatum Thunb.	Aragenatsuhaze
V. hirtum Thunb.	Usunoki
V. japonicium Miq.	Akushiba
V. oldhamii Miq.	Natsuhaze
V. praestans Lamb.	Iwatsutsuji
V. sieboldii Miq.	Honaganatsuhaze
V. shikokianum Nakai	Marubausugo
V. smallii A. Gray	Oobasunoki
V. uliginosum Linn.	Kuromamenoki
V. vitis-idaea Linn.	Kokemomo
V. wrightii A. Gray	Giima
V. yatabei Makino	Himeusunoki
V. yakushimense Makino	Akushibamodoki

* According to Makino (1970), Ohwi and Kitagawa (1983), and Uehara (1990).

These wild species, however, have not been improved through breeding, and only grow wild. Fruits of some wild species (*V. uliginosum* L., *V. vitis-idaea* L., *V. oldhamii* Miq. and *V. bracteatum* Thunb., etc.) were collected for use in making jam and liqueur.

Hara (1952, 1953) and Koike (1974) reported on *V. uliginosum*

L., and Kushima (1986) reported on five wild *Vaccinium* species native in Kagoshima prefecture. The plant form, size of leaves and berries were very variable in different locations and from plant to plant.

DEVELOPMENT OF BLUEBERRY CULTURE IN JAPAN

Blueberry cultivars were first introduced into Japan from the USA in 1951 (Iwagaki et al., 1977; Ishikawa, 1981). Since then, blueberry culture has gradually increased, and the growing area was about 190 hectare and the fruit production was about 400,000 kg in 1991 (Fruit Tree and Flower Culture Sec., 1993) (Figure 1).

The development of blueberry culture in Japan was due to the concentrated efforts of Dr. Iwagaki (previously Professor at Tokyo University of Agric. and Technol.). He studied the production of both rabbiteye and highbush blueberries. Many results of his studies (Iwagaki and Ishikawa, 1984), especially on propagation, fruiting, and marketing, contributed significantly to making today's blueberry industry possible in Japan.

Few evaluations of local adaptability were made in the early 1970s. After that, many researchers (Nakajima, Yokota, Tamada, Nakajo, Shimamura, Koike, Yokomoto and Kushima), all publishing in 1984, started trials on blueberry growing in local areas. As a result of these trials, blueberry growing expanded more rapidly beginning in the 1980s.

BLUEBERRY GROWING REGIONS IN JAPAN

Blueberry growing regions may be roughly divided into three zones in the Japanese islands, based on air temperature. (1) Highbush blueberries are grown in the northern parts and the highland areas of the main island that is relatively cool during the growing season (April-September); leading growing prefectures are Hokkaido, Aomori, Iwate and the highland areas of Nagano and Kumamoto. (2) Both highbush and rabbiteye blueberries are grown in the central region of the main island; leading prefectures are Gumma,

FIGURE 1. Changes in blueberry growing area and fruit production in Japan (according to Fruit Tree and Flower Culture Sec., Ministry of Agric., Forestry and Fishers., 1993).

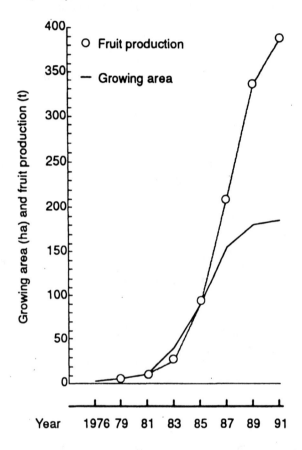

Chiba and around Tokyo. (3) Mainly rabbiteye blueberries are grown in the southern areas that are warmer during the growing season; the main growing prefectures are Hiroshima and Kagoshima.

THE DEVELOPMENT OF PICK-YOUR-OWN ORCHARDS

Blueberry production is divided into three types, namely production for the fresh market, for the processing industries and for pick-your-own farm sales (PYO). In recent years, PYO has been developing in the city and suburbs. For example, both highbush and rabbiteye blueberries are planted in the same field in Chiba prefecture. The harvesting season begins at the middle of June and lasts to the end of August. The entrance fee was about $5.00 per adult in 1993.

CULTIVAR AND VARIETY TESTS IN JAPAN

All of the highbush and rabbiteye blueberry cultivars in Japan have been introduced from the USA and other foreign countries. Leading cultivars of highbush blueberry are 'Weymouth,' 'Collins,' 'Blueray,' 'Bluecrop,' 'Berkley,' 'Dixi' and 'Herbert.' There are only three main cultivars of rabbiteye blueberry: 'Wooderd,' 'Homebell,' and 'Tifblue.'

Few cultivar tests were conducted in the early years of 1970. Therefore, many blueberry fields were planted without checking cultivar characteristics. After that, Ishikawa etc. (1979), Kushima (1981) and Yokota (1989) reported on the characteristics of many rabbiteye and highbush cultivars under local conditions.

Shimura's book (Shimura et al., 1993) described the characteristics of highbush and rabbiteye cultivars. It is the first book edited systematically in Japan, and is expected to be the most useful guidebook for the selection of cultivars.

Thirty cultivars of highbush and 20 cultivars of rabbiteye blueberries have been investigated at Chibaken Agric. College (Tamada, 1993a).

CLIMATIC AND SOIL CONDITIONS

Climatic conditions. The temperature and rainfall is very different from one blueberry growing region to another. The average temperature during the growing season (April-September) is at 21.1°C (70°F), and 5.4°C (41.5°F) in winter (December-February) at Tokyo where both highbush and rabbiteye blueberries are grown (National Astronomical Observatory, 1993) (Figure 2).

FIGURE 2. Average monthly temperature and precipitation at Tokyo and New York (USA) (according to National Astronomical Observatory, 1993).

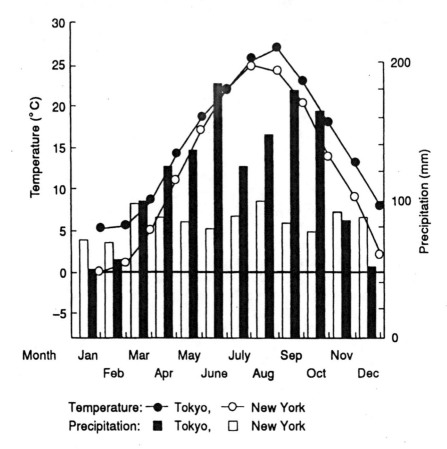

The average annual rainfall in Tokyo registers about 1500 mm (60 in). Rainfall of June and July is more than in other months. The harvesting period of highbush blueberries in Tokyo and the central parts of the Japanese islands is during this two month rainy season.

This is a very severe problem of highbush blueberry production. Ripening and harvest of highbush blueberries during the rainy season results in fruit quality that is often unsatisfactory. For example, the content of soluble solids is low, acidity is high, and shelf life is short (Tamada, 1989a, 1993a). Moreover, special care must be taken to insure that the blueberry roots (fibrous and shallow) are not injured by drought in the dry season.

Research is necessary to develop method or select cultivars suitable for a rainy season harvest. Another approach will be to establish protected cultivation for highbush blueberries (Ishikawa and Sugawara, 1993; Tamada, 1993a).

Soil conditions. Blueberries are grown on three types of soils in Japan. These are brown forest soils, red and yellow soils, and volcanic ash soils. The physical properties and chemistry of these soils are very different. However, the methods of soil management, irrigation and rates of fertilizer application for the different soil types have not yet been established in Japan.

PLANTING

Nursery stock. According to Japanese reports (Iwagaki and Ishikawa, 1984), highbush and rabbiteye nursery plants were produced mainly from hardwood cuttings.

Planting. Plantings are generally made with 2 year-old plants. The planting distance is generally about 1.5 × 2.5 m (5 × 8 ft) for highbush, and 2.0 × 2.5-3.0 m (6.5 × 8-10 ft) for rabbiteye blueberries. The planting hole is about 80-100 cm (32-40 in) in diameter, and 50 cm (20 in) in depth. About 30-50 L (31.5-52.5 qts) per hole of organic matter, such as peat-moss and chaff, is mixed with the soil to improve the soil acidity and soil aeration.

In the unique case of blueberry culture in Japan, both highbush and rabbiteye blueberries are planted in fields that had been rice paddies. Compared with the ordinary upland soils, these soils are very heavy, have little air space and poor drainage. If the drainage is not improved, the growth of blueberry plants will be very poor.

TRAINING AND PRUNING

The training and pruning are practiced according to the recommendations of Brightwell (1962), Eck and Childers (1966), Eck (1988) and Eck et al. (1989).

Two or three flushes of shoot growth occur during the growing season in the central and southern part of Japan. The flush of first growth continues until about the end of June (rainy season in the Tokyo area). June and July are the months when highbush fruit ripens, and when enlargement of rabbiteye fruit occurs. After this season, second and third flushes occur and very strong shoot growth develops every year.

It is desirable to establish the appropriate method of training and pruning that will decrease the competition of shoot growth with fruit enlargement.

FRUITING MANAGEMENT

Flower-bud differentiation. The flower-bud differentiation period is from the end of June to the beginning of September for highbush, and from the middle of June to the end of September for rabbiteye at Chiba prefecture; however, the cultivar and type of shoot affect the time (Tamada, 1984b). The flowering season lasts from the beginning of April to the beginning of May.

Necessity for cross-pollination. More than 80 percent fruit set must occur for a full crop (Eck and Childers, 1966). In Japan, research on fruiting of blueberry has been conducted (Tamada, 1977, 1991). It has become clear that cross-pollination usually increases the percentage fruit set, increases fruit size and causes earlier ripening compared with self-pollination of rabbiteye blueberries. The large size fruit had more seeds than the small fruit in rabbiteye blueberries (Kushima and Austin, 1979).

It was shown that the germination percentage of pollen was different from cultivars in rabbiteye and highbush blueberries (Tamada and Ishino, 1992).

SOIL MANAGEMENT

Mulching. Considering the characteristics of blueberry roots, mulching is necessary for favorable growth, and is especially beneficial for highbush blueberry (Eck and Childers, 1966; Cain and

Eck, 1966; Eck, 1988; Eck et al., 1989). In Japan, rice straw and chaff are available and are easily obtained by blueberry growers.

Water control. Judging from the amount of monthly rainfall, dry periods occur at late harvest time (from the end of July to August) and in late winter (in February). Most blueberry growers in Chiba prefecture have some kind of irrigation facilities, and they usually irrigate 5-10 L (5.25-10.5 qts) of water per plant at intervals of about three days.

Soil acidity. Controlling soil acidity is very important for blueberry growing (Eck and Childers, 1966; Cain and Eck, 1966; Eck, 1988; Eck et al., 1989). If the pH 5.5 or higher in rabbiteye, or higher than pH 5.0 in highbush, powdered sulfur should be mixed with the soil.

FERTILIZER APPLICATION

Recommended method of fertilizer application. Fertilizer sources, ratio of fertilizer sources, rates and time of application are important considerations in blueberry culture (Cain and Eck, 1966; Eck and Childers, 1966; Eck, 1988; Eck et al., 1989). The recommended method of fertilizer application must be adjusted for local condition (climate and soil), type of blueberry (highbush or rabbiteye) and cultivar. In general, 1-1-1 ratio fertilizers are applied to both rabbiteye and highbush blueberry, irrespective of the soil type in Japan (Tamada, 1984a).

Nitrogen nutrition. Blueberry plants are considered to favor the ammonium source of nitrogen and an acid soil (Cain and Eck, 1966; Eck and Childers, 1966; Eck, 1988; Eck et al. 1989).

Many studies of nitrogen nutrition were performed to investigate different forms of applied nitrogen (Tamada, 1984b, 1993b; Katakura and Yokomizo, 1987; Takamizo and Sugiyama, 1990, 1991; Sugiyama and Hanawa, 1992; Sugiyama and Hirooka, 1993). Applied rates (Tamada, 1984b), combination of nitrogen and other nutrient elements (Tamada, 1984b), combination of nitrogen and pH levels (Katakura and Yokomizo, 1988; Sugiyama et al., 1989) have also been investigated. From these studies it was concluded that rabbiteye and highbush blueberry growth was better with ammonium nitrogen than with nitrate nitrogen.

-

In a series of nitrogen source experiments (Tamada, 1984b), it was found that ammonium nitrogen (NH_4-N) resulted in good growth, and that nitrate nitrogen (NO_3-N) alone caused remarkably poor growth. Thus, nitrogen fertilizer should be composed of full or at least half ammonium nitrogen.

Diagnosis of nutrient condition. According to blueberry culture extended into the local, basic data is needed on what is good plant growth and fruit yield. Tamada (1989b) and Tamada et al. (1994ab) have been investigated nutrient deficiencies and nutrient excesses.

PEST CONTROL

Insects and diseases. The most common insects are the hair caterpillar, leaf folder and bagworm at the blueberry field of Chiba-ken Agri. College. Fungicides have been sprayed often on fields of highbush and rabbiteye blueberries. Some large scale blueberry growers use a spray calendar.

Birds. Starlings and sparrows are common fruit-eating birds during the harvesting season. It is necessary to cage the plants securely enough to prevent bird damage. A permanent frame is constructed over the bushes to support the netting.

HARVESTING AND MARKETING

Growth cycle of blueberry fruit. The growth cycle of blueberry fruits was determined to be a double-sigmoid curve (Shimura et al., 1986; Tamada et al., 1988). The growth cycle was divided into three stages. Stage I, young fruit stage, was the period of rapid growth after flowering. Stage II, retarded growth stage, was the period of little increase in fruit size. Stage III, maximum growth stage, was a second period that expanded until fruit ripened.

It has been shown that fruit treated with Ethrel matured earlier than nontreated fruit in rabbiteye blueberry (Fukushima and Gemma, 1990).

Harvesting. The harvesting time of the early cultivars in highbush blueberry ('Weymouth') begins at the beginning of June, and that of rabbiteye blueberry ('Woodard') at the beginning of July in the central part of Japan (Ishikawa et al., 1979; Tamada, 1984a; Fukushima et al., 1983, 1988).

The average amount of harvested fruits from mature plants is 4-6 kg (8.8-13.2 lbs) per plant in highbush and 5-10 kg (11-22.2 lbs) per plant in the case of rabbiteye blueberry. Four to 7 pickings are required at an interval of 5 to 7 days.

Much time and labor is required for hand harvest. No mechanical harvesters are used in Japan.

Marketing. The domestic blueberry production was about 400,000 kg (880,000 lbs) in 1991. The blueberry fruit is used primarily for processing rather than for fresh table use. Only about one-third or one-forth of the total domestic production is shipped to the fresh market.

Lowbush blueberry fruits are imported from abroad, mainly from Canada and the USA. More than 1,600,000 kg (3,520,000 lbs) are imported as frozen fruit (Society of Food Marketing, 1993) and are processed mainly into blueberry jam, cake, wine, etc.

Utilization. The studies of utilization and preservation on blueberry fruit are few in Japan. Ito (1984, 1990) gave details for the preservation and functional property of blueberry fruits. Takanami et al. (1980) and Hiroyasu et al. (1984) have researched the sugar, organic acid and free amino acid content of rabbiteye and highbush blueberries.

FUTURE OF BLUEBERRIES IN JAPAN

The future of the blueberry industry in Japan will depend on the production of high quality table fruit that most Japanese people prefer. The requirements are large fruit size, light-blue color, fine fruit shape, high content of soluble solids, low acidity, and superior keeping quality.

LITERATURE CITED

Brightwell, W. T. 1962. Rabbiteye blueberries (mimeograph). Coastal plain Exp. Station, Univ. of Georgia, USA.

Cain, J. C. and P. Lick. 1966. Blueberry and cranberry. In N. F. Childers (eds.). Fruit Nutrition. pp. 101-129. Somerset Press, Somerville, NJ, USA.

Eck, P. and N. F. Childers. 1966. Blueberry culture (eds.). pp. 3-378. New Brunswick, Rutgers Univ. Press, NJ, USA.

Eck, P. 1988. Blueberry science. pp. 3-220. New Brunswick, Rutgers Univ. Press, NJ, USA.

Eck, P., R. E. Gough, I. V. Hall and J. M. Spiers. 1989. Blueberry management. In G. J. Galletta and D. G. Himelrick (eds.). Small fruit crop management. pp. 274-332. Prentice Hall, Inc., Englewood Cliffs, NJ, USA.

Fruit tree and Flower Culture Sec, Agri. Production B, Ministry of Agriculture, Forestry and Fishery. 1993. Research on the fruit tree culture in 1991 year. pp. 47 and 105-106. (In Japanese).

Fukushima, M. and C. Oogaki. 1983. Studies on cultivation of rabbiteye blueberry (*Vaccinium ashei* Reade) in Tsukuba. Sogo Nogaku. 31(1):5-19. (In Japanese with English summary).

Fukushima, M., H. Gemma and C. Oogaki. 1988. Possibility of long term blueberry harvesting with combined highbush and rabbiteye blueberry cultivation in Japan. Acta Horticulturae 241:244-249.

Fukushima, M. and H. Gemma. 1990. Effects of Ethrel treatment on the harvesting date and harvesting period of rabbiteye blueberry. Bull. Agri. & For. Res. Univ. Tsukuba 2:87-93. (In Japanese with English summary).

Hara H. 1952. *Vaccinium uliginosum* L. in Japan. with references to variations in widespread northern species (1). Journ. Jap. Bot. Vol. 27(10):309-315. (In Japanese with English summary).

Hara H. 1953. *Vaccinium uliginosum* L. in Japan. with references to variations in widespread northern species (2). Journ. Jap. Bot. Vol. 28(3):83-92 (In Japanese with English summary).

Hiroyasu, T., H. Ishii, K. Takayanagi and T. Tamada. 1984. Studies on the sugars, organic acid and free amino acids in the rabbiteye blueberries. Tech. Bull. Fac. Hort. Chiba Univ. 33:1-5. (In Japanese with English abstract).

Ishikawa, S., T. Tamada and H. Iwagaki. 1979. Results of the performance of rabbiteye blueberry in Tokyo. Bull. Farms Tokyo Univ. Agri. Tech. 9:39-50. (In Japanese with English summary).

Ishikawa, S. 1981. The present status of blueberry growing and research in Japan. Koryo. 142:37-44. (In Japanese with English summary).

Ishikawa, S. and E. Sugawara. 1993. The experimental results with highbush blueberry under vinyl house. Acta Horticulturae 346:155-161.

Ito, S. 1984. Utilization and preservation of blueberry fruits. In Iwagaki and Ishikawa (eds.). Culture of blueberry. pp. 148-162. Seibundo Shinkosha. Tokyo. (In Japanese).

Ito, S. 1990. Functional property of blueberries. In Ito ed. Science of Fruits. pp. 125-129. Asakura shoten. Tokyo. (In Japanese).

Iwagaki, H., S. Ishikawa, T. Tamada and H. Koike. 1977. The present status of blueberry work and wild *Vaccinium* species in Japan. Acta Horticulturae. 61:331-334.

Iwagaki, H. and S. Ishikawa. 1984. (eds.). Culture of blueberry. pp. 1-239. Seibundo Shinkosha. Tokyo. (In Japanese).

Katakura, Y. and H. Yokomizo. 1987. Effects of pH level on growth and nutrients absorption of highbush and rabbiteye blueberry trees. Bull. Keisen Women's Junior College. 19:3-21. (In Japanese with English summary).

Katakura, Y. and H. Yokomizo. 1988. Effects of nitrogen form and pH of nutrient solution on the fractions of nitrogenous compounds in blueberry trees. Bull. Keisen Women's Junior College. 22:3-17. (In Japanese with English summary).

Koike, H. and K. Miyagawa. 1974. Studies on the blueberries. (1) On characteristics and propagations of the wild blueberry (*Vaccinium uliginosum*) in Japan. Nagano Hort. Expt. Sta. Rpt. 11:1-10. (In Japanese with English summary).

Koike, H. 1984. Blueberry growing in Nagano prefecture. In Iwagaki and Ishikawa (eds.). Culture of blueberry. pp. 206-221. Seibundo Shinkosha. Tokyo. (In Japanese).

Kushima, T. arid M. E. Austin. 1979. Seed number and size in rabbiteye blueberry fruit. HortScience. 14(6):721-723.

Kushima, T. 1981. Major varieties of rabbiteye blueberry, *Vaccinium ashei* Reade in Georgia State, U.S.A. Bull. Exp. Farm Fac. Agri. Kagoshima Univ. 6:1-29. (In Japanese with English summary).

Kushima, T. 1984. Blueberry growing in Kyushu district. In Iwagaki and Ishikawa (eds.). Culture of blueberry. pp. 229-235. Seibundo Shinkosha. Tokyo. (In Japanese).

Kushima, T. 1986. Survey of wild *Vaccinium* in Satsuma peninsula, Yakushima island and Okinawa island. Bull. Exp. Farm Fac. Agri. Kagoshima Univ. 11:19-29. (In Japanese with English summary).

Makino, T. 1970. Makino's new illustrated flora of Japan. p.467-469. Hokuryu-kan. Tokyo. (In Japanese).

Nakajima, F. 1984. Blueberry growing in Hokkaido. In Iwagaki and Ishikawa (eds.) Culture of blueberry. pp. 164-175. Seibundo Shinkosha. Tokyo. (In Japanese).

Nakajo, T. 1984. Blueberry growing in Gumma prefecture. In Iwagaki and Ishikawa (eds.). Culture of blueberry. pp. 201-205. Seibundo Shinkosha. Tokyo. (In Japanese).

National Astronomical Observatory. 1993. (ed.). Rika nenpyo :191-365. Maruzen. Tokyo. (In Japanese).

Ohwi, J. and M. Kitagawa. 1983. New flora of Japan. pp. 1172-1178. Shibundo. Tokyo. (In Japanese).

Shimamura, H. 1984. Blueberry growing in Tokyo area. In Iwagaki and Ishikawa (eds.). Culture of blueberry. pp. 187-192. Seibundo Shinkosha. Tokyo. (In Japanese).

Shimura, I., M. Kobayashi and S. Ishikawa. 1986. Characteristics of fruit growth and development in highbush and rabbiteye blueberries (*Vaccinium corymbosum* L. and *V. ashei* Reade) and the differences among their cultivars. J. Japan. Soc. Hort. Sci. 55(1):46-50. (In Japanese with English summary).

Shimura, I., K. Yokota, N. Hakoda, T. Kushima, S. Ishikawa, T. Tamada, H. Koike, H. Shimamura and I. Hagiwara. 1993. Investigation on the classification and varietal characteristics in the year 1992 (Blueberry). Horti. Labo. Tokyo Univ. Agri. Tech. (In Japanese).

Society of Food Marketing. 1993. Food, production, importation and consumption in the year 1993. p. 123. Society of Food Marketing. Tokyo. (In Japanese).

Sugiyama, N., I. Tanaka and T. Takamizo. 1989. Effects of pH and N form on the development of chlorosis in rabbiteye blueberry. J. Japan. Soc. Hort. Sci. 58(1):63-67. (In Japanese with English summary).

Sugiyama, N. and S. Hanawa. 1992. Growth responses of rabbiteye blueberry plants to N forms at constant pH in solution culture. J. Japan. Soc. Hort. Sci. 61(1):25-29. (In English).

Sugiyama, N. and M. Hirooka. 1993. Uptake of ammonium-nitrogen by blueberry plants. J. Plant Nutrition. 16(10):1975-1981. (In English).

Takamizo, T. and N. Sugiyama. 1990. Effects of N forms on plant growth and fractions of leaf N in rabbiteye blueberries. J. Japan. Soc. Hort. Sci. 58(4):865-869. (In Japanese with English summary).

Takamizo, T. and N. Sugiyama. 1991. Growth responses to N forms in rabbiteye and highbush blueberries. J. Japan. Soc. Hort. Sci. 60(1):41-45. (In English).

Takanami, S,. Y. Shinha, T. Yoshida and T. Nakajima. 1980. Free amino acid and organic acid composition of various blueberries. Kanzumejiho. 59(1):1-5. (In Japanese with English summary).

Tamada, T., H. Iwagaki and S. Ishikawa. 1977. The pollination of rabbiteye blueberries in Tokyo. Acta Horticulturae. 61:335-341.

Tamada, T. 1984a. Blueberry growing in Chiba prefecture. In Iwagaki and Ishikawa (eds.). Culture of blueberry. pp. 193-200. Seibundo Shinkosha. Tokyo. (In Japanese).

Tamada, T. 1984b. Fruit set, fertilizer application. In Iwagaki and Ishikawa (eds.). Culture of blueberry. pp. 83-87, 100-108. Seibundo Shinkosha. Tokyo. (In Japanese).

Tamada, T., M. Shinozuka and H. Kawashima. 1988. Growth cycles of fruit, and seasonal changes on the sugar content and acidity of blueberry fruits. Bull. Chibaken Agri. College 4:1-8. (In Japanese with English summary).

Tamada, T. 1989a. Blueberry growing in Chibaken, Japan. Acta Horticulturae: 241:64-70.

Tamada, T. 1989b. Nutrient deficiencies of rabbiteye and highbush blueberries. Acta Horticulturae 241:132-138.

Tamada, T. and M. Kihara. 1991. Effects of pollen parent on the fruit set, berry weight and number of seed per berry in highbush and rabbiteye blueberries. Bull. Chibaken Agri. College. 5:17-27. (In Japanese with English summary).

Tamada, T. and K. Ishino. 1992. Investigation of the optimum conditions for germination substratum of pollens, and varietal differences of germination percentage in rabbiteye and highbush blueberries. Bull. Chibaken Agri. College. 6:25-37. (In Japanese with English summary).

Tamada, T. 1993a. Some problems and the possibility of future on the blueberry industry in Japan. Acta Horticulturae 346:33-40.

Tamada, T. 1993b. Effects of the nitrogen sources on the growth of rabbiteye blueberry under soil culture. Acta Horticulturae 346:207-213.

Tamada, T., S. Nishikawa, Y. Yamakoshi and K. Iida. 1994a. Seasonal distribution of mineral elements in the leaf of blueberries, and investigation of leaf analysis of rabbiteye blueberry garden at some places in Chiba prefecture. Bull. Chiba-ken Agri. College. 7: (In press).

Tamada, T., T. Ishigami, M. Koshikawa, H. Saitou, T. Miyama. 1994b. Macronutrient deficiencies in rabbiteye blueberry. Bull. Chibaken Agri. College. 7: (In press).

Uehara, K. 1990. Encyclopaedia of trees with illustration. Vol. 3:525-539. Ariake-shobou. Tokyo. (In Japanese).

Yokomoto, M. 1984. Blueberry growing in Hiroshima prefecture. In Iwagaki and Ishikawa (eds.). Culture of blueberry. pp. 222-228. Seibundo Shinkosha. Tokyo. (In Japanese).

Yokota, K. 1984. Blueberry growing in Iwate prefecture. In Iwagaki and Ishikawa (eds.). Culture of blueberry. pp. 176-186. Seibundo Shinkosha. Tokyo. (In Japanese).

Yokota, K. 1989. Characteristics and selections of advantageous blueberry cultivars in Iwate. J. Fac. Agri. Iwate Univ. 19(3):149-159. (In Japanese with English summary).

List of Attendees

Ballington, James
Box 7609 Hort. Sci. Dept.
North Carolina State Univ.
Raleigh, NC 27695-7609

Bogash, Steve
18330 Keedysville Road
Keedysville, MD 21756

Braswell, John
P.O. Box 193
Poplarville, MS 39470

Brewster, Vickie
Cranberry & Blueberry Res. Ctr.
Rutgers University
Chatsworth, NJ 08019

Briggs, Bruce
Briggs Nursery, Inc.
4407 Henderson Blvd.
Olympia, WA 98501

Cline, Bill
NCSU/HCRS
3800 Castle Hayne Rd.
Castle Hayne, NC 28429

Darnell, Rebecca
Hort. Sciences Dept.
University of Florida
Gainesville, FL 32611

Draper, Arlen
640 E. Park Drive
Payson, AZ 85541

Ehlenfeldt, Mark
Blueberry & Cranberry Res. Ctr.
Rutgers University
Chatsworth, NJ 08019

Erb, Alan
Kansas State Univ.
Hort. Res. Ctr.
1901. E. 95th St. South
Wichita, KS 67233

Gallahorn, Charles
Maniilaq Association
P.O. Box 256
Kotzbue, AK 99752

Galletta, Gene
10300 Baltimore Ave.
Beltsville, MD 20705

Goulart, Barbara
Dept. of Horticulture
Penn State University
University Park, PA 16802

Gupton, Creighton
P.O. Box 287
Poplarville, MS 39470

Hancock, Jim
Dept. of Horticulture
Michigan State University
East Lansing, MI 48824

Hanson, Eric
338 PSSB
Dept. of Hort. MSU
East Lansing, MI 48824

Hillman, Brad
Dept. of Plant Path.
Rutgers Univ.-Cook College
New Brunswick, NJ 08903

Johnson, James
Dept. of Entomology
243 Natural Sciences Bldg.
East Lansing, MI 48824-1115

Korcak, Ronald
10300 Baltimore Ave.
Beltsville, MD 20705

Krewer, Gerard
P.O. Box 1209
Tifton, GA 31793

Kriegel, Robert
Dept. of Entomology
243 Natural Sciences Bldg.
East Lansing, MI 48824-1115

Lyrene, Paul
1137 Fifield Hall
Gainesville, FL 32611

Mainland, Charles
3800 Castle Hayne Rd.
Castle Hayne, NC 28429

Magee, Jim
USDA-ARS
Small Fruit Res. Sta.
P.O. Box 287
Poplarville, MS 39470

Nelson, Christopher
Maniilaq Association
P.O. Box 256
Kotzebue, AK 99752

Nelson, John
76683 16th Ave.
South Haven, MI 49090

Pliszka, Kazimierz
Warsaw Agric. Univ. SGGW
02-7660 Warsaw
Poland

Polavarapu, Sridhar
Rutgers Blueberry & Cranberry
Res. Center
Lake Oswego Rd.
Chatsworth, NJ 08019

Reich, Malinde
Manillaq Association
P.O. Box 256
Kotzebue, AK 99752

Rowland, Jeannie
10300 Baltimore Ave.
Beltsville, MD 20705

Scheerens, Joe
OARDC/Horticulture
1680 Madison Ave.
Wooster, OH 44691

Smagula, John
5722 Deering Hall-U. of ME
Orono, ME 04469-5722

Smith, Barbara
USDA-ARS Small Fruit Res.
P.O. Box 287
Poplarville, MS 39470

Spiers, James
USDA-ARS
Small Fruit Res. Sta.
P.O. Box 287
Poplarville, MS 39470

Stretch, Allan
USDA-ARS Blueberry
& Cranberry Res. Cent.
Lake Oswego Rd.
Hcol-Box 33
Chatswoth, NJ 08019

Takeda, Fumi
Appalachian Fruit Res. Cent.
45 Wiltshire Rd.
Kearneysville, WV 25430

Tamada, Takato
Chibaken Agricultural College
1059 Ienko, Toogane-shi
Chiba-ken 283
Japan

Trinka, Dave
MBG Marketing
P.O. Drawer B
Grand Junction, MI 49056

Vorsa, Nicholi
Cranberry & Blueberry Res. Ctr.
Rutgers Univ.
Chatsworth, NJ 08019

Walsh, Chris
Dept. of Hort.
Univ. of MD
College Park, MD 20742

Wildung, David
1861 E. Highway 169
Univ. of MN
North Central Expt.
Grand Rapids, MN 55744

Williamson, Jeffery
Fruit Crops Dept.
2113 Fifield Hall
Univ. of FL
Gainesville, FL 32611

Yarborough, David
5722 Deering Hall
Univ. of ME
Orono, ME 04464-5722

Zimmerman, Richard
10300 Baltimore Ave.
Beltsville, MD 20705

Printed in the United States
140655LV00002B/16/P

9 781560 220534